THE SANITATION, WATER SUPPLY AND SEWAGE DISPOSAL OF COUNTRY HOUSES

The Sanitation, Water Supply and Sewage Disposal of Country Houses

By

WM. PAUL GERHARD, C. E.

HONORARY DOCTOR OF ENGINEERING

Fredonia Books
Amsterdam, The Netherlands

The Sanitation, Water Supply and Sewage Disposal
of Country Houses

by
William Paul Gerhard

ISBN: 1-4101-0515-6

Reprinted from the 1914 edition

Fredonia Books
Amsterdam, The Netherlands
http://www.fredoniabooks.com

In order to make original editions of historical works
available to scholars at an economical price, this
facsimile of the original edition of 1914 is
reproduced from the best available copy and has
been digitally enhanced to improve legibility, but the
text remains unaltered to retain historical
authenticity.

PREFACE.

THE first part of the book treats of the general sanitation of country houses, brings a comparison of life in the city and in the country from a health point of view, dwells on the advantages of country life, and gives a condensed summary of the essential requirements of healthfulness in country houses. The soil, the subsoil, surface drainage, aspect, healthful surroundings and those which are objectionable, the cellar of the house, the lighting, heating and ventilation, the water supply, sewerage and plumbing, are briefly discussed.

In the second part detailed advice is given as to how to procure a satisfactory water supply. The sources of water, the various modes of raising it, the storage in reservoirs, elevated tanks or underground pressure tanks, and finally water distribution, are dwelt on at length and illustrated by actual examples from the author's engineering practice. This part of the book is the outcome of a number of lectures given recently by the author before the Civil Engineering Section and the Engineering Society of the Massachusetts Institute of Technology.

The third part discusses the all-important question of the sewage disposal for houses not in reach of sewers. Particular attention is given to the latest development in disposal methods, to the so-called biological or bacterial purification systems, including septic tanks, cultivation tanks, contact filter beds and sprinkling or trickling filters.

The illustrations accompanying the descriptions of examples of water supply and sewage disposal are largely taken from the author's practice. The author does not consider it necessary to offer an apology for doing this, as the reasons are sufficiently self-evident; but the statement is made in order to anticipate criticism on the part of reviewers. At the suggestion of his publishers, however, a few examples of water supply and sewage disposal plans from the practice of other engineers and engineering contractors have been added. In the list of illustrations the source from which they are taken is given.

Throughout the book the aim of the author has been to inform and tell his readers "WHAT TO DO" rather than "How to do it." In other words, he has endeavored to establish leading sanitary principles which in turn should lead to correct sanitary practice. He thought it best to omit from the book many details regarding the execution of sanitary methods, thereby perhaps avoiding the encouragement of "amateur sanitary engineering," which is quite frequently met with nowadays, and which generally leads to dismal failures.

THE AUTHOR.

33 Union Square, New York.
February, 1909.

CONTENTS

I. SANITATION

II. WATER SUPPLY

Examples of Water Supply Systems

III. SEWAGE DISPOSAL

Examples of Sewage Disposal Systems for Country Houses

LIST OF ILLUSTRATIONS

I.

SANITATION

SANITATION OF COUNTRY HOUSES

WHOEVER has the opportunity of planning and building for himself a home in the country should bear in
mind the importance of having not only an attractive
and comfortably arranged house — be it a cottage or a
mansion — but, above all else, a healthful house. Upon
the healthfulness of the home will chiefly depend the
comfort, well-being, and happiness of its inmates.

A country home is beautiful, not only in summer but
at all seasons, and life in the country implies, as a
rule, the existence of health-favoring natural conditions.
Hence it is all the more important that the building
itself, its surroundings, its water supply, drainage, and
sewage disposal, in a word, all its vital household
arrangements, should be quite sanitary.

Essentials of Sanitation in Dwellings. — In a paper on
". Sanitary Engineering" published in the initial number
of the *American Architect*, issued on January 1, 1876,
the late Colonel George E. Waring, Jr., an eminent
authority in sanitary engineering, wrote as follows
regarding the essential requirements of sanitation for
dwellings:

"In the interest of sound building it is of the utmost
importance that those who are charged with the construction
of houses, especially for residence, should concern themselves not only with matters of taste, and of economical and
substantial construction, but with the fulfillment of the chief
(because the first) purpose of all building, that of providing

protection against influences injurious to health. The first use of the house was doubtless to furnish shelter against exposure, whose influence would be permanently or temporarily injurious to the physical condition for those for whom it was prepared. The mere provision of shelter has so long been accomplished that this consideration no longer enters into the conscious idea of those who build.

But there has arisen within a comparatively short time, as an incident to the adoption of modern household conveniences, the knowledge of other influences to which the human frame is subject, which equally demands the wisest and most careful attention.

Every house of any pretension is provided, as a matter of course, with certain arrangements for the removal of its liquid wastes, which it has been too often the custom of architects to treat simply as matters of specification, and over which they exercised less personal supervision than was given to more conspicuous parts of their work."

In a paper on "Sanitation of Interiors of Houses" read at the International Congress of Hygiene held in Paris, in August, 1900, it was held that "all sanitarians are agreed now that pure water, pure air and free and abundant light are the three great essential factors indispensable to the salubrity of a house, and that it follows from this that one of the principal objects of the sanitary engineer should be to supply the house with pure water and salubrious air."

While agreeing in the main with this statement, I am inclined to enlarge on it somewhat, for in my judgment, in order to be healthful, a dwelling should fulfill the following essential requirements :

1. It should be dry and free from dampness, be well drained, and have the ground air carefully excluded from it.

2. It should be well sewered, so that all liquid foul matters are nowhere stagnating but are immediately, speedily and completely removed.

3. It should have a sanitary system of plumbing pipes, fixtures, and fittings.

4. It should be well lighted, and have as much sun exposure as possible.

5. It should have a copious supply of pure water, flowing under a good pressure and protected in the house from all sources of possible contamination.

6. It should have proper and sufficient means for ventilation, and be supplied with pure air in all its parts.

7. It should be kept clean and free from any nuisances.

8. It should have healthful surroundings.

Disadvantages of Town Houses. — Happy the man who is the owner of a healthful house in healthful country surroundings! With the modern tendency towards concentration of a large population in cities, people in moderate circumstances have to content themselves in these with living in "flats" or apartments, which are narrow, crowded, gloomy, generally ill-arranged, badly ventilated, and which offer little, if any, of the privacy which the smallest cottage home in the suburbs offers. The poorer classes are huddled together in wretched tenement houses, sometimes so wrongly called the "homes" of the poor. But even the wealthy citizen, unless he be a real Crœsus, lives in town in a four-story house, which has a narrow frontage, with only a few windows, extends far to the rear of the lot, and which stands in the midst of a row of similar houses. Such city houses can doubtless be made prominent and perhaps attractive in outside appearance by giving them a richly ornated façade, but they have at best only a small yard space for air and light in the rear, while

on the two long sides they are enclosed by other buildings, or possibly they adjoin one of the sky-scraping apartment houses which are towering up everywhere, depriving
the streets and the lower buildings of air and sunlight.

An architectural writer applied the term "stupid city
homes" to such town dwelling-houses. Bad as their
outside often is, their interior is many times worse.
The majority of city houses, unless built within the
past decade, have radical sanitary defects, which may
be the contributory cause of preventable illness and
premature death, and which swell thereby the mortality
rate of cities.

Fortunately, however, this is the era of "rapid tran-
sit," and the available and ever-increasing means of easy
communication, by rail and by boat, offer facilities for
living not too far from the great centers of business,
and of acquiring an easily accessible home, located at a
sufficient distance from the heart of a great city. Thus
people who are fond of a quiet rural life but whose business compels them to go to the city, may enjoy the
advantages and comforts of a suburban or country
home, located in a good neighborhood, and in healthful
surroundings.

Suburban and Country Houses. — In the following pages
I offer some general suggestions on domestic sanitation
which should be of interest and of value to people of
moderate means who have wisely resolved to reside in a
detached house in a suburb or in the country. Further
on in the book the question of water supply for country
houses and estates is taken up more in detail, while
the third part discusses the all-important problem of
the disposal of the sewage of all those buildings which

cannot avail themselves of a connection with a city sewer.

Among the natural advantages, which a residence in the country implies, are the following: There is no over-crowding, and no noise from horse-cars, railroads, or factories; there are no dirty streets and no noxious smells; instead, one can breathe pure air and enjoy plenty of sunlight and blue skies; one has also number-less wild flowers, wide, green fields and meadows, numerous shade trees and fruit orchards; there is plenty of opportunity for healthful exercise; and instead of the never-ceasing noise and turmoil of the busy streets of a metropolis, one may hear the song of birds and the delightful murmuring of brooks, while the eye may rest on the many beauties of the landscape, on lakes, forests, hills, meadows, or the ocean.

Those living in houses already built may not find it too late to consider and possibly follow some of my suggestions, while those about to build will do well to study them with a view of adapting them to their special conditions. Even those, whose "ideal home" at present exists only in their dreams, should, when they decide to commence building operations, carefully medi-tate over the advice contained in these paragraphs, and endeavor to apply it with profit when the proper time arrives.

The character of the site and of the general surround-ings necessarily has a great influence upon the develop-ment of the house plans. Consequently, the first step toward building should be, not the planning of the house, but the selection of a salubrious site. In choos-ing a lot, due thought should be given to all sanitary considerations. After the building lot is selected, the architect may proceed with the floor plans, and here again, in planning, proper attention should be paid to the sanitary arrangements. False economy with regard to them always, sooner or later, leads to annoying and often serious consequences.

Among sanitary defects that ought to be studiously avoided are those relating to the site or locality, to the surroundings, the aspect, and to the character of the soil and of the subsoil. While it is possible to improve defects of construction, an error in the location, in the prospect or aspect of the house is beyond remedy. Unhealthy conditions found to exist as regards the site, the prospect, and aspect should therefore always lead to the abandonment of a proposed site.

But, given a healthful site and desirable surroundings, we must study further how to keep the ground, upon which the house is erected, and the soil around the dwelling, free from dampness and from accumulation of organic matter. We must also avoid defective construction, such as would cause dampness of walls, wet cellars, smoky chimneys, rapid decay of building materials, etc. Furthermore, we must arrange the floor plans, not only with reference to comfort and convenience, but also with regard to aspect, healthfulness, proper exposure to sunlight, ample space in rooms, large amount of window surface, proper height of ceilings, etc. Intimately connected herewith is the question of light and ventilation, and proper means for the effective removal of fouled air and for the introduction of pure air of a proper temperature. Next comes the problem of how to warm our house in the most healthful and economical manner. Again, we have to provide for an ample supply of pure water for all domestic purposes, and have to select with care and judgment the plumbing fixtures of the house, and last, but not least, we must devise proper means for the removal and disposal of the household wastes, which should be accomplished wherever possible, with due regard to the utilization of the refuse materials for agricultural purposes.

Location and Site. — The site for a house should receive the most careful attention. First of all, the site should be dry. Those who build homes in the suburbs or in the country are generally not very restricted in their selection, and hence may avoid low lands, and the neighborhood of swamps, marshes, stagnant ponds, mill dams, and polluted creeks. These localities are very apt to be damp, chilly, and may breed malarious fevers. A location on or near the banks of a river, which is subject to periodical flooding, or which has a sluggish flow, cannot be considered healthy. Neither should a house stand in a ravine or valley. High ground is naturally preferred in selecting a building site. But the top of a hill would, in many cases, expose a house unnecessarily to strong and often cold winds. The best and by far the most pleasant location is on a hillside, or on a knoll, facing the south or southwest. It is not advisable, however, to build a house too close into a rather steep hillside, for a building so located will be damp and generally unhealthy, owing to the insufficient circulation of air around it. As regards trees, it may be said that, while it is true that they are pleasant neighbors in affording shade in hot weather against the sun, and shelter against raw winds, they must not stand too near a dwelling. If so, they darken the rooms, prevent the entrance of sunlight, deprive the house of proper currents of air, and promote dampness of the walls and of the cellar.

In choosing a site for a country home, always remember the importance of an ample and pure supply of water for domestic purposes. Wherever a supply from a well is contemplated, have a few preliminary borings made to ascertain the amount and character of the water which the proposed well will yield. Borings will, at the same time, afford valuable information with regard to the level of the ground water.

If the lot appears wet, examine carefully into the best manner of draining it. If no outlet for the subsoil drains seems available — although this is not often the case — avoid the site.

Inquire into the direction of the prevailing winds, and note if your site affords any shelter against them. You may plant a screen of evergreens which will afford such a shelter, even in winter time, at the same time adding to the beauty of your surroundings.

Surroundings. — It is important that proper and searching attention should be paid to the neighborhood where one intends to build. A noisy factory, a tannery, soap works, rendering establishments, or a railroad station are not pleasant neighbors. If the adjoining lots are built upon, it is worth while to inspect the buildings and their immediate surroundings. If the neighbor has a stable in the rear of his house, or perhaps a privy or leaching cesspool, the site should be avoided, unless there is reason to believe that he can be persuaded to appreciate the importance of good sanitary surroundings. Unless he feels disposed to cooperate in such matters as drainage and sewerage, the future will bring constant annoyances arising from nuisances in the neighbor's back-yard, no matter how thoroughly one may care for one's own sanitary arrangements. However effectually, for instance, a well may be guarded against contamination from drains or soilpipes, its water may become tainted and ultimately cause disease by a hidden connection or soakage from the neighbor's privy, or from his ill-constructed leaky drains.

Consequently, where land is cheap, it is always best to build a small cottage on a large lot; this cuts off and isolates the house from constant annoyances such as I have mentioned, besides offering a good chance to plan and arrange a small kitchen or vegetable garden; it will also secure air and sunlight and afford better means for a proper refuse disposal. Always place the dwelling at a distance from the roadway, for this avoids the dust in summer, and gives opportunity for planting shade trees and shrubs. One more word of advice in regard to the site. Where the side of a hill is chosen, examine most scrutinously into the drainage of any houses built on the hill slope above your site.

Sunlight and Aspect. — It is desirable that every dwelling should be so situated as to receive direct sunlight and pure outside air from all four sides; in other words, houses to be truly cheerful and salubrious should stand *free* and *detached* from one another. This will, no doubt, increase the annual coal bill, but it is also sure to decrease the annual doctor's bills. Human beings need the cheering, purifying, invigorating influence of the light and warmth of the sun, no less than plants do. Therefore, houses should be so placed, with regard to the four principal directions of the compass, that each side may receive some direct rays of the sun for at least a few hours of the day.

A true northern aspect is bleak and cheerless. A southern or southeastern aspect is considered the best with us, for in winter time it is less cold and chilly, owing to the warmth of the sun, while in summer time it receives the cooling southern breezes. In this matter of aspect people who build country houses enjoy advantages which are not available in a city, for the above-mentioned desirable features can often be obtained by a slight change in the location of the house. It is not, moreover, necessary to place a country house parallel to the street or road passing in front of it. On the contrary, it generally adds much to the picturesque appearance of a cottage to have it standing irregularly on the lot, especially if the walks leading to it and the lawn or flower beds are laid out with a certain regard to tasteful landscape architecture.

Character of Soil. — In selecting the site for a house, a loose, porous soil is preferable to ground liable to be damp or wet. Pure, dry sand and gravel make excellent sites for building purposes. Next to these, rocky soils may be chosen, which are, as a rule, quite healthful. Clay soils, which are more or less impervious to water, and therefore always damp and chilly, and alluvial lands, must not be chosen as a site for dwellings. But, above all, avoid *made* land. Although this restriction applies more particularly to building lots in towns, it is not uncommon, even in the suburbs of large cities, to find low ground filled with garbage, rubbish and decaying vegetable and animal débris, which are prime causes of impure air in dwellings. A virgin soil or ground which has not before been built upon is, undoubtedly, preferable to sites of old, torn-down buildings. If the latter must be taken, a detailed and thorough examination should be made with respect to the purity of the soil. Some lots are literally honeycombed with cesspools, privy-holes, or have a network of broken drains full of accumulated filth, and the soil is at times found to be contaminated from liquid house refuse, or by the soakage from barn-yards and stables. A well should never be sunk through such formerly occupied ground. It is quite important to ascertain, by preliminary borings, the level of the ground water, for a high water level means continuous dampness, and must be abated by thorough under-drainage.

Subsoil Drainage. — The under-drainage of a site effects a permanent lowering of the ground level, and thus secures to a proposed dwelling dry foundation walls, and absence of dampness from the house interior. To remove the subsoil water, small porous, round tile-drains, two to three inches in diameter, should be laid with open joints at least two feet below the level of the cellar floor. The general arrangement of the lines may vary somewhat in each case, but ordinarily the branch drains can be laid in parallel lines, their distance varying from ten to twenty-five feet, according to the amount of water to be removed. Wherever springs are found, special lines may be required. The trenches should be refilled with broken stones or coarse gravel. The branch pipes should be collected into a larger main pipe, and for this a three or four-inch tile pipe will answer in most cases. This main drain should be continued with proper fall to a ditch, ravine, or water course. There must never be any connection between the subsoil drains and any foul-water drain or sewer, or with a cesspool or sewage tank.

If the dwelling stands on a hillside, exposed to subsoil water flowing over an impervious stratum, the side walls of the house nearest to the hill are very apt to be wet, often so much so as to have the subsoil water percolate through the cellar walls. In this case, the subterranean water vein should be cut off by a blind drain, i.e., a trench dug above the house sufficiently deep and carried with proper fall diagonally across the lot. The trench must be filled with broken stones and carried down the hill to some outlet, either an open ditch or a brook.

Removal of Surface Water. — Attention should be paid to the proper removal of surface water. In the case of suburban cottages the rain falling upon the roof is very frequently collected and stored for use in underground cisterns. Occasionally a public water supply is available, and in such case the cistern is omitted, and the roof water is allowed to run away on the surface. A large part of it soaks into the ground and thereby tends to keep the foundation walls damp and unhealthy. To avoid this evil, the grounds surrounding the house must be properly graded, in order to shed the water away from the walls. At points fairly remote from the house the surface water may be permitted to soak away into the ground, and the vegetation will help to absorb a part of it. In other cases, however, surface channels or gutters must be arranged, especially where there is a clay soil.

Ground Air. — Besides water, the upper layers of the soil always contain ground air. This has a tendency to rise into the house, especially in winter when heated dwellings act as huge chimneys, drawing up large quantities of air from the ground beneath them. Such exhalations, which consist in the case of a pure soil of carbonic acid and watery vapor, but which in the case of a contaminated soil are largely mixed with gases of decomposing organic matter, should be rigidly excluded from the interior of houses. For this reason, dwellings without a cellar should never be placed immediately on the ground, but should be placed on piers, arches, or posts, being thus raised sufficiently to allow of a large air space and perfect circulation between the ground surface and the floor beams. This will, at the same time, prevent the quick rotting of the joists and floor boards. In order to avoid a cold basement, its flooring should be laid double, and an intermediate space provided, to be filled with a nonconducting material, such as mineral wool.

Cellar Floors. — Although somewhat more expensive, it is advisable to excavate for a cellar and to build the house on strong, well-made foundation walls. The floor of the cellar must be made perfectly tight against ground water and ground air. There are different ways of doing this. One of the best methods is the following : Cover the surface of the cellar, which has previously been levelled, with a layer of large broken stones and on top of these put concrete, at least four or, better, six inches deep. Next put on a thin layer (about one-quarter inch) of hot, pure asphaltum, and on top of this a finishing coat, one inch thick, of Portland cement.

Dampness of Cellar Walls. — Cellar walls must always be made impervious to dampness. As usually built, they are extremely porous, and moisture rises in them by contact with the adjoining ground and by capillary attraction. The best plan to prevent dampness of walls is to have a complete cut-off between the foundation walls and the ground, by an open area, carried completely around the building. The area must be well drained and ventilated. This form of construction is, however, expensive, and a similar isolation may be accomplished by building double or hollow walls, the space between the inner and outer walls being well aired. The foundation walls should be placed upon a bed of concrete, and be covered on their outside with a layer of asphaltum to a point somewhat above the level of the ground. It is very important to provide, at this height in the wall, an isolating or damp-proof course, which may consist of a horizontal, thick layer of asphaltum, of slate, bedded in cement, of layers of tarred roofing paper, or else of hollow tiles. The sill and the floor joists must, of course, be kept above the damp-proof course. The surface water may be kept away from the outer walls by filling the space next to the wall, to a depth well below the foundation walls, with broken stones or gravel. Sometimes a tile-drain is placed *below* the footing course to carry off any accumulation of percolating storm water. This trench may be covered at the top with a stone slab to shed off surface water.

The so-called "practical" builder will probably sneer at some of these suggestions. I can assure those of my readers who care to build a *healthful* home, that the money spent for such preventive measures will form an excellent investment. The proper construction of healthy foundation walls, and of a cellar which is dry and cheerful at all times, is the basis of sanitation in cottage-building. This much accomplished, all remaining requirements are not so difficult to fulfill.

Light and Air in Cellars. — Next to dryness, the most desirable features of a good cellar are, that it should be well lighted and perfectly ventilated. Good light in a cellar helps much toward keeping it in a proper condition. It is advisable to carry a ventilation flue down to the cellar and to have an opening into it near the cellar ceiling. It is not difficult to appreciate the importance and necessity of cellar ventilation, if we remember that floors necessarily have some crevices or shrinkage holes, through which the cellar air will rise and mingle with the atmosphere of the living and sleeping rooms. Above all other things, do not allow your cellar to be made a sort of gigantic poke-hole for rags, cast-off clothing, old shoes, tin cans, rotten vegetables, garbage, swill, or other offensive matters. See that it is kept at all times free from rats and vermin. Do not tolerate any opening in the cellar floor for the removal of surplus water into foul water drains. Such opening, even if trapped, will be sure to act at times as an inlet for unwelcome sewer air.

Cellars often become damp in summer time from condensation of moisture from the air. Contrary to popular notion, it is best, during hot weather and when the outside air is muggy and damp, not to open the cellar windows during the day. The cellar is so much cooler than the outside moisture-laden air, that the watery vapor condenses on the walls and floors of the cellar, producing dampness and causing mildew and rust. In such weather the cellar windows should be opened only during the night.

The cellar should contain besides the heating apparatus the bins for the fuel used in heating and for the kitchen range. It is also the place for the main drainage pipes, and for the main supply services for water and for gas, including the gas meter. A cool place with an even temperature should be partitioned off for a wine cellar. If the cellar is quite dry, parts of it may be utilized as a storage place, and it is best in that case to provide a raised wooden platform.

Arrangement of Rooms. — Due attention should be given, in grouping the rooms of a house, to the question of aspect or outlook. For this reason it is a great mistake to draw the plans of a house first, to select the site afterwards, and then to try and make the latter conform to the plans. Given the lot, the various rooms should be arranged as much as possible with regard to the outlook. Living rooms should front towards the south or southeast; the principal bedroom may have an outlook towards the east or northeast, thus enjoying the morning sun's rays. A dining-room may look towards the north, northeast, or northwest, while the domestic quarters will usually be located on the west or northwest side of the dwelling. Of course, these rules cannot always be strictly adhered to in the case of the

smaller cottages, and they are given merely as suggestions, to be followed where practical.

A house should in general be so placed as to get the
greatest amount of sunlight to the interior. All rooms
should be airy, sunny, and well lighted. Nothing is so
detrimental to domestic cleanliness as darkness. Dark
staircases and closets are an abomination. Every room
of the house should have large, good-sized, outside windows, reaching well up to the ceiling. Roofs of wide
porches or piazzas are delightful sheltering places against
the scorching heat of an August sun, but they rob
the lower rooms of much necessary sunlight. Shut-
ters and blinds are desirable things to keep out too
much sun, but they must not be kept closed all day as
is — alas ! — the custom with so many people. Householders may promote the interests of their pocketbooks
by preventing the early fading of curtains and carpets,
but their ultimate object is generally lamentably defeated
by an increase in doctors' bills, caused by the continued
ill-health of members of the household who are spending
the greater part of the day at home.

The excellent rule of Bacon that *"Houses are built
to live in, not to look upon"* should constantly be kept
in mind in arranging the house interior. Select the
largest and most cheerful rooms for the bedrooms.
Avoid placing a sleeping room on the ground floor of a
dwelling, and never use the basement for such purpose.
Next in importance comes the living room, in some
houses the study or library, which should face the
south, southeast, or southwest. If one can afford to
have, in addition to the sitting room, a reception room
or parlor, this should be the smaller of the two. The

dining-room should be located either next to the kitchen
or preferably should communicate with it (where the
size of the dwelling admits this) through a butler's
pantry. The kitchen should be large, roomy, light, and
airy, for this facilitates domestic operations. Wherever
possible, windows should be arranged on opposite sides
of a kitchen in order to have cross-ventilation in summer time. If the size of the building lot and the sum
available for building admit of it, a large hall should
form an integral part of the house. It should have a
cheerful outlook upon the landscape, and be fitted up
in a cosy manner so as to permit its use as a reception
or sitting room.

Low ceilings — though considered fashionable by
many — are not conducive to health. No room ought
to be less than nine feet in height. Windows ought to
reach nearly to the ceiling, and should open at top and
bottom. Any stagnation of air at the ceiling may then
be avoided by lowering the top sash. For bedrooms
the cubic space for each person should not be less than
one thousand cubic feet.

**Some Details of Sanitary Construction, Materials, Wall Sur-
faces, Floors, Windows, and Furnishings.**— I do not intend
giving rules which may be found in treatises on building
construction, but will briefly mention a few special
points, worthy of careful attention in building a salubrious dwelling.

Moderate-priced cottages are naturally constructed
largely or wholly of wood, with the exception of the
cellar walls, which are generally of stone. Healthful
dwellings may be constructed either of stone or brick-
work, or else of framework, but all materials must be

careful.y selected and proper use must be made of them
in the construction. Frame dwellings, when situated
where they are much exposed to driving rains, are apt
to be damp and affected by mildew, and their woodwork
may rot sooner.

In the class of dwellings under consideration, the
inside of the walls is usually plastered ; in like manner
partition walls are constructed of wood and then lathed
and plastered. A sanitary wall-surface should be smooth
and non-absorbent, so that it could be washed clean
with soap and brush, such as, for instance, a tiled wall
or one made of glazed bricks. Partition walls should
be impervious in order to prevent air currents passing
from one room to another, or from one floor to the
next above. When built impervious, they also check
the rapidity of a fire.

Outside walls need not be absolutely impervious, and
the porosity of outer walls, as usually built, helps to
some extent in changing the air of a room, as shown by
Professor Pettenkofer. It seems quite doubtful, therefore, whether there is wisdom in the practice of covering the outside walls of buildings with some absolutely
impervious coating.

Plaster is necessarily absorbent, consequently many
of the so-called organic vapors or impurities of the air
gradually penetrate the wall. Hence arises the peculiar
musty smell often found in rooms of old buildings.
Wall papers are not much better in this respect, but
painted or whitewashed walls are probably the best.
Regarding wall papers, attention should be called to
the danger from arsenical poisoning, existing not only
in the *green* colors, but also in a large variety of other

tints. The paste used to fasten wall papers often decomposes, and gives rise to annoying odors.

The usual flooring in cottages and suburban dwellings consists of a board floor, laid with more or less wide floor boards. As a rule, it is carelessly jointed, and this is very objectionable because it favors the accumulation of dust and dirt under the floor. It is customary, though hardly wise, to cover all floors, even those in bedrooms, with carpets cut to the shape of the room and tacked down to the boards. But carpets used in such a manner are neither artistic nor healthful. They are harboring places for vermin, and accumulate a large amount of dirt and filth. They are especially objectionable on this account in bedrooms, as well as in dining-rooms, and too much praise cannot be given to the rapidly spreading fashion of using loose rugs in place of carpets.

Hard-wood floors should be laid with close joints, preferably tongued and grooved; the strips should be narrow, and well-dried hard wood only should be selected. In place of carpets, small and large rugs should be used, and the uncovered outer edges of the floor should be painted, varnished or waxed and polished. If cost is no objection, a wood carpet or parquet flooring may be used. In cleaning the floors of a house, the use of much water should be avoided, for some of it necessarily soaks through the cracks in the floor boards, and the moisture which remains, together with the dust and dirt which settle into these places, cannot help being detrimental to health.

The dark days of the Middle Ages — dark in more senses than one — are fortunately past, when a tax was

considered necessary upon the number of windows which
a house had. Yet one sees even nowadays houses
where builders seem to have been afraid of such a
"window-tax." All rooms of a house should be well
lighted by large windows opening at the top and the
bottom, or by so-called casement windows, the latter
used almost exclusively on the continent of Europe.
There should be a certain minimum amount of window
surface in proportion to the cubic space, or better the
floor space, of a room. A common rule is to have
one square foot of window surface to every one hundred
cubic feet contents of a room. The proportion between
window surface and floor space in houses is put by various sanitary authorities at from one-seventh to one-
twelfth.

Regarding sanitary furnishings, I quote from Dr. B.
W. Richardson:

> "In furnishing, woolen and fluffy materials are bad, heavy
> curtains to beds and windows are bad, carpets which
> cover the whole of a room are bad. In a word, *all
> materials that catch dust, keep dust, hide dust, and, on
> being shaken, yield clouds of dust, are bad.*"

In summer time especially, a house should be free
from heavy hangings and from upholstered furniture
and decoration which accumulate dust and dirt.

Window and Door Screens. — The windows and doors
of country houses should be screened during, at least, a
part of the year, in order to prevent the entrance of flies,
mosquitoes, and other insects.

In the better class of houses the screens are fitted
permanently to the windows and consist of wooden or
steel frames, holding a screen of very fine mesh. The

material for the wire screens is either black or painted iron, or, better, brass or copper. The screens may be fitted to the windows either inside or outside of the sashes. An advantage of the outside position of the screen is that the window sashes may be raised or lowered without handling the screens, but the drawback is that they are more liable to corrosion, being always exposed to the rain and the weather.

For less expensive houses adjustable and removable screens may be used, and in the case of cheap dwellings the windows may be screened with cotton gauze or netting.

Screens are particularly desirable for the bedrooms of the house and also for the dining-room, the kitchen, and the pantry. They serve to increase our comfort during sleeping hours, because they spare us the annoyance of being bitten or waked in the early morning hours. In the dining-room they help to protect the food on the table, and in the kitchen and pantry they have for a long time been considered almost indispensable, because without them it would be impossible to prevent the fly nuisance and the aggregation of hundreds of these pests near or on the materials from which our food is prepared.

The question of convenience and comfort is, however, not the only one with which we are dealing when we attempt to screen our windows and doors. Experiments and investigations made in recent years have established the fact that both the flies and the mosquitoes are carriers of disease, and with this in view we must consider the screening of our houses as a *sanitary measure* of infinite importance, as will be explained further on.

As regards flies, it is not so very long ago that the modern view has taken the place of notions that flies are harmless or even of some service to mankind. Even in the 1901 edition of Johnson's (Appleton's) Universal Cyclopedia an article on "Flies" says: "Many species of flies are to be regarded as beneficial, as they *act as scavengers* and *remove much noisome matter*."

Flies and Mosquitoes as Carriers of Disease. — That the bite or sting of mosquitoes, flies, and other winged insects may produce illness, blood poisoning, or even death has been known for a long time. In recent years, however, the very important discovery has been made that these insects play an important rôle in the dissemination of bacteria and in the transmission of pathogenic germs.

Flies. — The common house fly cannot bite, and the transmission of disease takes place in a different manner. The flies breed principally in stable and privy manure. They enter our houses and are attracted by the food in the kitchen and in the dining-room. Excrementitious matter and bacilli attach themselves to the bodies and to the feet of the flies and are in this way transferred to our solid food or to the milk; that a large amount of typhoid fever is thus transmitted has been demonstrated almost beyond the shadow of a doubt, and it was particularly the experience in the United States military camps during the Spanish-American War of 1898, which clearly pointed to the flies as the spreaders of the typhoid fever. A commission of army surgeons investigated the matter thoroughly and in a voluminous report presented detailed evidence to show that the flies were the most active agents in the spread of this disease. From their report I quote as follows:

"The latrines contained fæcal matter specifically infected with the typhoid bacillus. Flies alternately visited and fed upon this and upon the food in the mess tents. More than once it happened, where lime had been scattered as a disinfectant in the pits, that the flies with their feet covered with lime were found afterwards walking over the food. The typhoid fever was much less frequent among members of messes who had their tents screened than it was among those who took no precaution."

Many years previous to this war, Dr. Joseph Leidy attributed the spread of gangrene in the Washington hospitals during the Civil War to the flies. Experiments made by English physicians in India seem to point to the fact that other diseases than typhoid fever, for instance, Asiatic cholera and the bubonic plague, may be transmitted by flies. In Africa the tsetse fly is the cause of the transmission of the much-dreaded cattle disease. Flies are also said to transmit the eye-disease known as "pink eye," and in Egypt the Egyptian eye disease.

Mosquitoes. — Mosquitoes are blood-sucking insects which play a very important rôle in the transmission of diseases, the chief of which are malaria and yellow fever. In both cases the mosquitoes act as intermediary host for the disease; in other words, it is necessary that the mosquitoes should first bite or sting a patient sick with yellow fever or with malaria. In doing this they suck up the germ of the disease, which develops in their blood, and when a certain time has elapsed the sting or bite of such an infected mosquito will cause the disease in a healthy human being.

It is an interesting fact that not all mosquitoes transmit disease, and that different diseases, such as malaria and yellow fever, are transmitted by the bite of different kinds of mosquitoes. Malaria, which up to a few years ago was attributed chiefly to the noxious exhalations of swampy regions, is transmitted by the bite of a single species, namely, the *anopheles*, and yellow fever is so far known to be transmitted only by a species, the *stegomyia fasciata*, which infests largely the Southern Atlantic coast, a part of the Southern States and the Islands of Cuba and Jamaica, as well as some of the smaller Antilles.

The transmission of malaria by mosquitoes was made the study of an Italian commission who experimented in the Roman Campagna, and in a similar way yellow fever was discovered to be due to the *stegomyia fasciata* mosquito by the United States Army Commission, who made important experiments and discoveries in Havana, Cuba, the results of which led to the almost entire extermination of the once so dreaded scourge.

The United States Army Board summarized the results of its experiments and observations regarding the transmission of yellow fever by mosquitoes as follows:—

First: "All attempts to bring about infection through contact of beddings, clothing and dejecta of yellow fever patients failed. The yellow fever is transmitted only by a mosquito of the species *stegomyia*.

Second: Yellow fever patients can be the source from which other cases originate only when they have been bitten by the proper mosquito. All cases and suspected cases should therefore be kept behind safe mosquito screens and netting.

Third: A yellow fever patient is dangerous only when bitten by a mosquito during the first three or four days.

Fourth: Hospitals for treatment of suspected cases of yellow fever should be located on high and well-drained grounds, away from creeks, pools, or standing water, free from mosquitoes and not surrounded by grass or shrubbery. All entrances and exits to hospitals should be provided with close mesh wire screen spring doors, and similar screens fixed immovably over every window, or any other opening communicating with the outer air.

Fifth: Standing water should not be permitted in barrels or vessels of any kind; broken crockery, tin cans, or other possible retainers of rain water should be searched for and removed.

Sixth: All surface pools should be promptly drained and filled in with gravel or covered with petroleum. Apply kerosene also to the standing water in ditches, pools, and rain water gutters. The margins of ponds should be deepened to enable fish to reach the mosquito larvæ.

Seventh: Water should not stand uncovered in houses; rain water in cisterns or barrels should be covered with petroleum if not used for drinking; if so used all vents and openings should be tightly screened or covered. Make a periodical search for the wigglers of mosquitoes, and in this way reduce their number and the chances of infection."

The ordinary layman has as yet very little understanding for the sanitary importance of fencing out flies and mosquitoes from the interior of our houses, and physicians are just beginning to appreciate the fact that to prevent the further spread of malaria, typhoid, or yellow fever it is absolutely necessary that the patient should be screened so that the insects may not reach him.

Practical Suggestions for the Extermination and Control of Mosquitoes and Flies. — *Flies*. — Besides the window and door screens already mentioned and the use of sticky fly paper, fly traps or similar devices for catching and destroying flies in the house, it is necessary that more radical efforts should be made to destroy the flies where they breed. First of all the manure from horse stables should be treated with lime or kerosene and kept in securely closed pits until it is removed, and this removal should be accomplished at frequent intervals. Inside of houses the maintenance of absolute cleanliness will be of much assistance in dealing with this evil. The use of kerosene or other oil has been suggested for privy vaults and cesspools. Dirt accumulations of any kind around a house should be abolished, and in this way much can be done to mitigate the evil. In cities the reduction in the number of horse stables and the increasing use of automobiles, bicycles, motor cycles, and electric trolleys help very materially in the fight against these pests.

Mosquitoes. — The use of window and door screens and of mosquito nettings over the beds is advisable, likewise the hunt for mosquitoes with cups filled with kerosene and the burning of pyrethrum powders in rooms. Of much more importance is the abolition of the breeding places of mosquitoes and the destruction of the larvæ or wigglers. All cisterns, rain water barrels, water tanks, and cesspools about the house should be well screened. All stagnant pools of water should, wherever possible, be drained or filled in, or else these places may be treated with kerosene oil. In ponds and pools of clear water the introduction of small fish, which eat up the larvæ, is recommended.

The great difficulty in the warfare against the mosquitoes lies in the fact that combined action is necessary in dealing with them. In many parts of this country, for instance on Long Island, on the Jersey Coast, and in other places, regular mosquito brigades have been organized by Associations or by the State Entomologists and have accomplished excellent results, particularly where large swampy areas have been ditched and drained. It is a fact worth noticing that all remedies and measures of prevention commend themselves to us as general sanitary measures of importance, such as, for instance, the maintenance of gutters in a clean condition and the prevention of accumulations of stagnant water in them, the deepening of ponds along their edges, and the improvement of the banks of running streams.

The American Mosquito Extermination Society have done a great deal of useful work in this direction, and in a "Mosquito Brief" recently issued they state: "Mosquitoes are a needless and dangerous pest; their propagation can be largely prevented by such methods as drainage or filling of wet areas, removal, emptying, or screening of water receptacles, and spraying standing water with oil where other remedies are impracticable. Attention should be paid to cisterns, house vases, cesspools, road basins, sewers, watering troughs, roof gutters, old tin cans, holes in trees, swamps, and puddles. As malarial mosquitoes may be bred in clear springs, the edges of such places should be kept clean and they should be stocked with small fish. The breeding and protection of insectivorous birds, such as swallows and martins, should be encouraged. Thorough screening of houses and cisterns is necessary to prevent the spread of malaria or yellow fever. The continued breeding of any kind of mosquitoes with the attendant menace to public health and to the life and comfort of man and beast is therefore the result of ignorance or neglect."

Much more might be said about health and comfort in the house, for no attempt has been made to cover all points. The few suggestions on the principal topics are made merely to emphasize the importance of this subject, but we must now turn our attention to matters of more immediate sanitary importance, which are comprised in what may be termed "domestic engineering," namely the lighting, ventilation, heating, water supply, plumbing, and sewage disposal of country houses.

Lighting. — Suburban and country houses are artificially lighted, either by individual, and usually portable, sources of light, such as oil lamps and candles, or else by some central system of lighting, having one common source for all rooms, halls, and other parts of the house, such as a gasoline or air-gas machine, an acetylene generator, or an electric lighting plant.

It is always advisable, while a building is in course of construction, to put in the gas pipes and the conduits for the electric wiring, even when no lighting plant is contemplated for immediate instalment. The underground gas mains of a nearby town may, at some time, be extended to the building lot, and an electric light and power company may in the future run its feeders near the house. In both cases, it will be an advantage to have the house already piped for gas and wired for electricity, for otherwise the process of tearing up floors and cutting walls and partitions to put in the gas pipes and the electric wiring conduits will entail not only a great deal of discomfort but likewise a heavy expense.

When the gas piping is put in, it should be tested, before the plastering is done, by a force pump and mercury gauge to make sure that it is perfectly tight and

that there are no concealed leaks. Equally desirable
and necessary is the examination and testing of the
electric installation by the underwriters, particularly
when the wiring for the lights is put in. In the case of
gas piping, large sizes of pipes should be chosen so that
they would be suitable and ample in case the owner
decides to put in a gas machine, for gasoline air gas
requires piping somewhat larger in size than does ordinary city gas.

The lighting of rooms by means of kerosene oil lamps
is a method comparable, as regards convenience, to the
individual heating of rooms by a number of separate stoves
or fireplaces. Compared with gas lighting fixtures, oil
lamps have the advantage of portability, but this very
advantage renders them more dangerous in use, for oil
lamps are liable to be upset, causing fires and damage to
property. Oil lamps are also liable to explode and become
a danger to life. They constitute a cheap and desirable
light, particularly for reading and for sewing, for which
occupations they are preferable to electric or gas lights.
But they have some drawbacks, such as the bad odor
often emitted, the heat given off by large oil lamps of
the duplex burner type, the smoking of lamps, and the
soiling of the hands in carrying them about. Then
again, if we consider the labor and expense involved in
cleaning and filling the lamps, in trimming and renewal
of the wicks, and the expense incident to the breaking
of lamp chimneys, gas light is about as cheap, particularly so in the larger towns where the price of gas has
been reduced. Explosions of gas are comparatively
rare, and in most cases are due to criminal carelessness
in searching for gas leaks with an open light, whereas

lamp explosions cannot always, even where great care is exercised, be foreseen or prevented.

The lighting of rooms with candles is almost entirely confined to emergency lighting, except possibly the lighting of the dining table by means of dainty wax or paraffine candles, held in graceful candlesticks or candelabras. The soft and mellow light of candles lends itself particularly to the adornment of the festive table, and to the decoration of drawing-rooms generally, but it is neither a convenient nor a cheap method.

In recent years some improved forms of individual acetylene lamps have been introduced, and a few types have even received the approval of fire underwriters. They give a brilliant white light and are inexpensive, both as to the first cost of the lamp, and cost of the carbide, but the body of the lamp must include a water reservoir, and this makes the lamp cumbersome and heavy. Some intelligence is also required in charging these lamps, and they cannot be intrusted to the care of the average domestic.

If a central system of lighting is wanted, owners of country houses may choose between

1. A gasoline gas machine.
2. An acetylene gas machine, or
3. An electric lighting plant.

Gasoline gas machines consist essentially of a tank or reservoir to hold the fluid, a blower to force the air over the gasoline and to generate pressure, a mechanism to run the blower, a mixing chamber and the required pipe connections. Gasoline is a fluid which has a pungent odor and the property of evaporating under ordinary temperatures. When it evaporates, it mixes with the air and becomes inflammable, and in certain proportions of air and vapor explosive. Hence it is somewhat dangerous to use, and great care is required where such machines are installed, particularly in the recharging or refilling of the generator. No fire, open light, or even a burning cigar should be used near the gasoline tank. Gasoline must never be stored inside of a building, and the rules of fire underwriters require that it be kept in an underground vault, or brick or stone building, placed at a distance of 50 feet from the dwelling-house.

The carburetted air-gas is a simple mechanical mixture of common air and of the vapor of gasoline ; it is heavier than common gas, and to obtain a more uniform quality of the gas it is passed first through a mixer or equalizer before it is delivered into the house gas pipes. This mixer is also intended to prevent the smoking of the gas flames.

The blower or air pump may be located in the cellar of the building. The blower is operated in various ways, either by a weight suspended from a drum, and wound up in a manner similar to a clock, or else a water wheel is used to run the air pump, but this type of machine can only be installed where the water supply is plentiful, and the water running to waste should, if possible, be utilized. At any rate, it is not advisable to discharge it into the house drain where the latter discharges into a sewage disposal field.

The outside generator and the air pump or blower are connected by means of tightly jointed iron pipes. The air pipe as well as the pipe carrying the carburetted gas must be pitched towards the generator in order to return to it all condensation. The blower draws its free air from outdoors, and not from the cellar.

The air gas can be used not only for lighting, but also for heating and for cooking. For light, it is best to burn the gas in incandescent mantle burners, which should be specially adjusted for this particular kind of service. Ordinary gas burners are not suitable, and even the special air-gas burners give, as a rule, a poor and unsteady light, and are very apt to smoke up the ceilings.

In recent years, acetylene gas machines have been introduced and used in some cases instead of carburetted air-gas machines. As is well known, acetylene gas is generated by the action of water on calcium carbide. The gas obtained is called acetylene gas, and it gives a very brilliant, white, and steady light. It is cheaper than electric light, requires but little attention and, if used with care, is perfectly safe.

There are numerous acetylene generators in the market, and house owners are advised to select only from those apparatus which have been approved by the underwriters.* Once the type has been decided upon, purchase a generator ample in size and capacity for the duty or service required. Of the various types of machines, those in which carbide is fed into the water are considered to be the best.

Some underwriters require the machine to be placed in a frostproof outbuilding, at some distance from the

* The National Board of Fire Underwiters, of Chicago, issue a list of approved acetylene generators, which list is frequently revised and brought up to date, and which is sent free on application.

house; others permit locating the generator in the basement or cellar. In this case it should be placed where there is good light for the inspection of the machine, and where it can be conveniently recharged.

In connecting the generator with the main gas riser of the house, the pipe should be given a good grade back to the machine. Before connections are made, make sure by a test that the gas piping is absolutely airtight, because acetylene gas mixed with air is highly inflammable, and even the slightest leak may become a source of danger.

It was formerly thought that since acetylene light was burned in very small burners, viz. one-half cubic foot burners, the piping could be made a great deal smaller than required for city gas. But this idea has proved a fallacy, and acetylene gas specialists now advise the use of larger pipes; for instance, twenty burners require a pipe not smaller than three-quarters of an inch.

Electric lighting of country houses from individual plants has many advantages. The light does not vitiate the air, nor does it give off much heat. The incandescent bulbs lend themselves more readily than gas jets to the decorative lighting effects sought. Electric lighting is also somewhat safer than gas or oil lamps as regards danger from fire, but it is so only if all rules and precautions advised by the underwriters are strictly followed and observed. The wires should never be run in wooden moldings, but must be carried in iron or brass "armored" conduits. The entire house installation must be tested by an expert electrician to guard against defects.

The plant required for the lighting by electricity may

be located in the barn or in the water pumping station, if there is one, and comprises an engine or power motor and a dynamo or generator. Water power may be utilized in turbine wheels to drive the dynamo; sometimes a steam engine is used, and the exhaust steam from the engine is then utilized during the cold weather to heat the house; in summer time it may be used to heat the water for the household. Gasoline and oil engines are employed more than other power motors; they have the great advantage of being capable of being used for other purposes during the daytime, for instance, for sawing wood logs, or for running the pump of the water supply plant.

The connection between the engine and the dynamo is either a direct one or else a belted connection. In some cases storage batteries are installed in connection with the dynamo, insuring a steadier supply of electric current.[*]

Ventilation. — Ventilation, or the change of air, must go on in dwellings at all seasons of the year. Its aim is to remove the vitiated air in a dwelling and to introduce a sufficient amount of pure air, moderately heated in winter time and supplied with a proper amount of moisture. This should be thoroughly and uniformly diffused in the house interior, in gentle currents, without causing any drafts. Drafts are dangerous to health, because they rob the human body too suddenly of a part of its heat. In summer time, ventilation is well and easily accomplished by opening doors or windows, and by an occasional "air flushing," i.e. by creating cross

[*] See *Gerhard*, The American Practice of Gas Piping and Gas Lighting in Buildings, 1908.

currents through rooms. Fireplaces should not be covered up in summer by boards. In winter time, ventilation should be combined with the heating of the house, and further suggestions about it will be given when speaking of the various modes of warming cottages.

In the spring and autumn months we often content ourselves with a small wood or coal fire in the open fireplace, and in such a case the easiest way to provide for incoming fresh air is by admitting the same through the windows, and directing the cold current to rise up to the ceiling. This may be done by lowering the upper sash and raising the lower one slightly, but not enough to leave openings at top or bottom. A better way is, of course, to have a ventilating open fireplace, such as the "fire-on-the-hearth" stove, or other similar apparatus.

The so-called spontaneous or accidental ventilation which goes on continually by reason of air penetrating walls cannot, practically, establish a sufficient change of air. Its effect is very much reduced by papering, painting or plastering of the wall surfaces, and by treating the outside of the building walls with some water-proof process, as is frequently done to obtain a protection against driving rainstorms.

Methods of Warming. — The question of how to warm a country house will depend upon the climate and locality of the dwelling, and to a great extent also upon its exposure. Three methods of warming the air of halls and rooms are available, namely, warming by open fireplaces, by stoves, and by hot-air furnaces. The methods of warming by direct or indirect radiation with steam or hot water apparatus, however good they are, cannot be considered here, as they require a larger out-

lay of money than is usually available for the buildings
under consideration.

Open Fireplaces. — Ordinary fireplaces warm principally by radiation, the heat from the fire being imparted
to surrounding objects or persons without warming the
surrounding air to any great extent.   The degree of
heat varies inversely with the square of the distance
from the grate fire.   Thus it happens in very cold
weather that with a fireplace as the only means of heating a room of an exposed dwelling, a person near the
fire may be almost roasted, while at the opposite extreme
end of the room the temperature may be almost down
to the freezing point. A further disadvantage is the
fact that an open fire warms only the part of the body
which faces the fire.   The greatest objection to the
ordinary open grate fire lies in the fact that 85% and
more of the fuel is wasted, the heat from it going
straight up the chimney flue.   A fireplace generally
causes extremely annoying drafts from window cracks,
or from spaces between the door and the trim or saddle,
especially in very cold weather.   On the other hand, if
the cracks of windows are carefully closed and stopped
up the chimney is apt to smoke, owing to a deficient air
supply.   While, therefore, an open fireplace may be
adequate in warm climates, it is entirely inadequate to
warm, *per se*, cottages in our eastern, northern, and
northwestern states.

To say that a very large waste of fuel occurs in
warming by fireplaces is not strictly correct. The heat
going up the flue is not actually quite *wasted*, for it
forms a good aid to the ventilation of rooms. We shall
see later that, as an accessory of other heating methods,

the fireplace is eminently serviceable, and much to be recommended.

Ventilating Fireplaces. — Better than ordinary fireplaces are the improved so-called ventilating fireplaces, which are provided with a large air-chamber, and a sufficient air supply from outdoors.   There are several excellent devices of this kind in the market, and these are, of course, much more economical as far as the burning of fuel is concerned, about 35% of the heat being utilized, while a good ventilation of the rooms is also accomplished.

Stoves. — In this country, stoves of cast iron and of wrought iron are the usual and most economical means of heating cottages and small suburban dwellings.   As ordinarily fitted up and arranged, stoves constitute the worst possible devices for warming the air of our rooms. Heating should always be combined with ventilation, that is, there should be a continuous removal of the fouled air and introduction of plenty of pure air arranged so as not to cause inconvenient or unhealthy drafts.   A room warmed by an air-tight stove soon contains air which has become entirely unfit to breathe, because a stove removes practically none of the vitiated air, and there is usually an entire absence of any provision for introducing fresh air.   It is true that less fuel is con- sumed, and stove heating is consequently economical, at least apparently so, but in reality it causes loss of strength and vigor, general debility and extreme sensi- tiveness, and hence increased doctors' bills.

If a dwelling must be heated by stoves, the following precautions should be observed.   Select a good-sized, well-built stove, with tight joints, one which is lined on the inside with fire-brick to prevent the iron from getting red hot and to retain, as much as possible, the heat. A supply of fresh, pure air from the outside should be provided and carried to a jacket surrounding the stove; in this way the pure outdoor air is warmed by contact with the stove, and then circulated in the room. The smoke-pipe of the stove should be large, and should not have a damper to shut off the draft. A valve may be placed on the fresh-air inlet pipe to regulate the amount of fresh air introduced at will. For the removal of the foul air, outlets leading into the chimney flue must be arranged near the ceiling of the room, care being taken to prevent down-drafts or the entrance of smoke, by arranging a self-closing flap valve at the outlet. It is much preferable, however, to have a separate exhaust or ventilating flue built in the chimney alongside of the smoke flue and warmed by the latter; this flue should have outlets into every room through which the chimney passes. The stove should have ample capacity to heat the room even in very cold weather without driving the fire to a red heat. It is a good plan to supply a moderate amount of moisture to the air of the room by placing a water kettle or evaporating pan on the stove.

Warm Air Furnaces. — Heating suburban and country dwellings by means of warm air furnaces has many advantages over stove heating. Furnace heating is, strictly speaking, stove heating, with this difference, however, that there is provided only *one* large stove, centrally located in the basement or cellar, from which air pipes of sufficient size carry the warm air into the rooms as desired. There is, consequently, much less labor in carrying coal and making fires, less trouble in keeping up the fire, and less dirt and dust from removing ashes.

Furnace heating is disliked by many, and some have even condemned it as being detrimental to health. While it is undoubtedly true that improperly arranged furnace apparatus may be the cause of ill-feeling, headache, or sickness, it is unreasonable to apply these objections to the system as such, for it is well known that heating by furnaces can be made perfectly healthful and agreeable. It is impossible to successfully heat a room by furnace heat unless arrangements are made, by an open fireplace or other outlet into a chimney flue, for the withdrawal of the air which has been exhaled and which is fouled by respiration of the occupants of the room. It is impossible to introduce into a room, pure, warmed air, without removing a like amount of fouled air. Another mistake, frequently made, is to take the air supply to the furnace air-chamber directly from the cellar. Thus, cellar air, ground air, or air from leaky sewer pipes is often sent up in a heated condition into the living and sleeping rooms of the house.

If warming by means of a furnace is decided upon, care should be taken to select from the innumerable patterns in the market a good furnace. The furnace should be of the best quality of material of its kind — either cast iron, wrought iron, or soapstone — and of a good size, for if the furnace is small, it will become overheated in extremely cold weather. This is very objectionable, because it renders the air less fit for breathing, and is liable to cause cracks in the cast iron, and to loosen the joints in wrought-iron furnaces. The furnace must be well constructed, the fire pot must be lined with fire-brick to prevent the rapid burning out of the iron, the joints must be few in number and perfectly tight, and this must be made the subject of a special examination.

Fresh Air Conduits. — The furnace should have either one or two large cold-air ducts, leading to the outside of the house, and located on opposite sides of the house, if there are two ducts. These air ducts should take their fresh air supply preferably from a point five or more feet above the surface of the ground. A slide-valve or damper must be arranged in the cold-air box, to regulate the amount of the in-coming air, and where there is danger from impurities in the air the air supply should be filtered through a loose cotton filter. At the mouth of the air box a wire netting should be placed to prevent the entrance of rats or other animals. The air box should be constructed of well-dried, wooden plank, with closely fitted joints. Better, although more expensive, is a galvanized sheet-iron air duct with locked joints. It is advisable to carry the cold-air box along the ceiling of the cellar, where it is in sight, and not below the ground, where it may and often does become filled with ground water or pools of sewage from broken cellar drains. The size of the fresh-air inlet should be approximately equal in area to the aggregate sum of the areas of all hot-air flues which lead from the air chamber of the furnace to the rooms. The fresh air should be kept tolerably moist by arranging in the air chamber of the furnace an evaporating pan kept constantly full of water.

Flues and Registers. — The furnace must be located as centrally as possible, so as to make the horizontal hot-air flues short, for the velocity of the air current is reduced by friction in long tubes, especially when these flues are small. The vertical hot-air flues should, preferably, be kept on the inside walls, and must be as direct as possible, and of ample capacity. The inlets or registers for admitting warm air into the room should not be placed in the floor. It is unhealthy to stand over them; moreover, floor registers form receptacles of dirt and dust, are unsightly in the floor, and combustible articles carelessly dropped into them may cause a fire. The inlets should be placed in a side wall, preferably six or seven feet above the floor. To avoid the danger of charring the woodwork, no hot-air flues should come in direct contact with floor joists, boards, or partitions; all woodwork should be securely protected by some non-conducting material. The smoke pipe should be large and be connected with a smooth flue of proper size, so as to insure a good, steady draft, which will remove all gases of combustion. There should be no damper on the smoke pipe, and the fire should be regulated only by more or less admission of air under the fire grate. Overheating of the furnace must be avoided, for this unduly dries the air, scorches the organic matter in the air which comes into contact with the fire, and thus causes a peculiar, disagreeable smell.

Ventilation by Means of Open Fireplaces. — An open fireplace in the hall and in the principal living and sleeping rooms constitutes, in connection with furnace heating, the most comfortable and pleasant arrangement for withdrawing the fouled air from the rooms. Where pure warmed air is introduced by heating registers, the radiant heat from a fireplace is particularly invigorating and comforting. The enjoyment of a cosy home is increased when its occupants can gather around a cheerful, glowing, open fire on the hearth to exchange pleasant thoughts or to dream away twilight hours in looking at the flickering light.

When open fireplaces are not provided for ventilation, outlets must be arranged for leading into ventilating flues, built parallel to the smoke flues of the chimneys. Chimney flues should preferably not be built against outside walls, for if placed in such a position they are apt not to draw well, unless a special air space is arranged at the outside of the flue to prevent its too rapid cooling. Ventilating flues must be built without sharp angles; they should be smooth on the inside and preferably round in section. Where they remove the air from a number of rooms, their cross-section must be proportionately increased. Bedrooms should never be heated by base burner stoves, but should have a fireplace acting at all times as an efficient foul-air flue. It is advisable to heat the principal halls moderately, in order to avoid cold drafts throughout the dwelling. The bathrooms and kitchens should be ventilated with special care.

Water Supply. — The following hints on water supply refer chiefly to the smaller suburban cottages and country dwellings. In the second part of the book the water supply of country buildings and estates is taken up at greater length, and the arrangements necessary to obtain a perfect supply system are described more in detail.

In the case of isolated country houses it is rare to find available a public water supply delivering the water under a suitable pressure. As a rule, each owner is compelled to provide his own supply, and the most common sources of water are either a spring, a well, or a rain water cistern.

Wells. — It is a common sight in the country to find a well located close to or adjoining a leaching cesspool or a privy. Usually these wells are shallow, being dug or sunk to a very limited depth, and frequently the liquid sewage from cesspools soaks through the porous subsoil down to the subterranean water stratum which is tapped by the well. The danger to health from drinking impure and polluted water is now universally acknowledged. Polluted well water is rendered more dangerous by the fact that it often has a bright, sparkling and clear appearance, and that in summer time it has a low temperature, making it particularly agreeable as a beverage. Only a chemical, microscopic and biological analysis can reveal its unwholesome condition. It is extremely difficult to fix a limit of minimum distance between a well and a cesspool or privy, as so many different factors have to be taken into consideration. In rocky ground, especially, there may exist hidden fissures carrying the contents of cesspools to a much greater distance than would generally be expected.

If there is no leaching cesspool, no privy, and no other cause of soil contamination in the neighborhood, a well may be safely used. But if cesspools must be kept on the ground surrounding the cottage, or if they exist in the neighbor's lot, and if such ground has been previously saturated with filth, a well should not be sunk.

A properly constructed well should have tightly built walls so as to be impervious from the level of the ground water up to the surface. In this way any infiltration of impure liquid from the upper soil surrounding the well may be prevented. The surface of the ground should be raised somewhat at the well, and graded so as to pitch in all directions away from the well. This prevents the entrance of surface washings. The opening of the well must be thoroughly covered, in order to prevent the falling into the well of vermin and smaller animals, or the washing in of decaying vegetable or organic matter.

Driven or Tube Wells. — Better than dug wells are those known as "driven wells," tube wells, or "Abyssinian" wells. They are constructed as follows: A wrought iron tube, $1\frac{1}{2}$ to 2 inches diameter, having at its lower end a steel point or shoe perforated with numerous holes, is driven into the ground, which must, of course, be free from stones or boulders, until the ground-water table is reached. If necessary, several lengths of tubing are screwed together by means of couplings, but for an ordinary suction pump the vertical depth of tubing, or the "suction-lift" should not exceed 25 to 28 feet. The upper end of the tube is attached to the pump, and continued suction will soon wash away the sand at the lower end of the pipe, and furnish a stream of clear water.

Rain Water Cisterns. — Wherever a well cannot be sunk, cottages should be supplied with rain water. This is collected from the roof and stored either in a tank placed in the garret, or else in an underground cistern. The latter keeps the temperature of the water moderately low throughout the year. Persons unaccustomed to drinking rain water object to it on account of its flat taste, but if it is carefully collected, properly stored, boiled before use or filtered and then cooled with ice and well aerated, it makes an exceedingly wholesome and agreeable drink.

To determine the amount of rain water available from a certain roof, ascertain the amount of surface of its horizontal projection, and multiply this by the annual rainfall in feet and decimals of a foot. The total amount in cubic feet must be divided by two, to allow for unavoidable loss through evaporation and for wasted, impure roof washings. It is easy to arrive at a proper size for the cistern, if the amount of water which can be made available is known or estimated.

In collecting roof water, it is important to allow the first washings from the roof, which always contain more or less organic filth in the shape of dust, horse dung from the street, excrements of birds, leaves from trees, etc., to run off on the surface. This may readily be accomplished by placing cut-offs on the rain-water pipes, which are worked by hand or else may be arranged to act automatically. The best roofing surface for collecting rain water is slate, and next to this shingles.

Underground cisterns are usually built circular in shape, of hard burnt brick, laid in hydraulic cement. The walls of the cistern must be made perfectly water-tight, not only to prevent leakage, but also to prevent the entrance into it of ground water from the outside. If an overflow pipe is provided, it should under no circumstances communicate with any drain or sewer, or discharge into a cesspool. As soon as delivered into the cistern, the water must be kept scrupulously clean, and any possible source of pollution should be removed.

It is a good plan to build into the cistern a filtering chamber to remove by straining the coarser impurities in the water. This may be done by building a partition wall in the cistern, thereby establishing a small clear water chamber, in which the suction pipe is placed. The dividing wall is built with courses of brick, some of which are laid dry, and thus act as strainers. This arrangement, it need hardly be said, wants periodical cleaning as much as any of the household filters. Cisterns should be frequently inspected, emptied, and cleaned ; the opening at the top must be closed by a solid cover, to prevent the falling in of vermin, mice, rats, etc., and to guard against contamination by surface washings.

Springs. — Dwelling-houses may also be supplied from a spring, by a gravity supply in case the spring is located on a hillside higher than the house, or if the spring is situated at a lower level than the house a hydraulic ram or a windmill may be used to pump the water. The spring should be enclosed by walls and a cover provided to guard the water against contamination. Sometimes the spring is arranged to supply a small collecting basin or reservoir, in which the night flow can be stored, where pumping is intermittent, as with a hot-air or internal-combustion engine.*

Filtration of Water. — The water supply for drinking purposes is often purified by means of domestic filtration. This is especially desirable in the case of cistern

* For a more extended discussion of the subject "Water Supply for Country Houses," the reader is referred to the author's pamphlet "The Water Supply of Country Houses," price 40 cents, postpaid; published by the author.

water. Household filters should act not only as strainers by removing suspended impurities, but they ought to act also chemically by oxidizing a part or all of the dissolved organic matter; most important of all, however, is the requirement that they should be germ-proof, i.e. retain the germs which the water may contain. Various materials are used for domestic filters, amongst them being sand, sponge, flannel, animal charcoal, spongy iron, porcelain, natural sand-stones, and compressed infusorial earth. Nothing is more erroneous than the supposition that a filter, once started, will continue to act forever, without further attention. Whatever the filtering material may be, it should be frequently cleaned, aerated, boiled, and sterilized; from time to time it should be entirely renewed. It must, therefore, always be easily accessible. The smaller filters, which are attached to the faucets on the supply pipe, are generally made reversible, and if this latter operation is regularly performed, they work quite well, although their action is of necessity largely a mechanical one only, i.e. they only strain the impurities of the water. One of the best germ-proof filters is the Berkefeld filter, which is made in several sizes, and so as to be attachable to the kitchen faucet. Even this filter requires cleaning about once a week, and from time to time it is advisable to sterilize it. Larger filters are connected by means of piping with the pressure supply, and these, too, answer well, provided they have proper washing arrangements, enabling the periodical reversing of the direction of the filtering current. Other household filters consist of portable vessels, filled by hand and not directly connected with the supply pipes. Filters are

also sometimes placed in cisterns or at the end of the
suction pipe leading to the well.

Service Pipes. — Pipes for conveying water to the
plumbing fixtures may be of drawn lead, of tin-lined
lead, or of block tin. Wrought iron is used extensively, either plain, galvanized, enameled, or made
rustless by the Bower-Barff process; rubber-coated,
glass-lined, and tin-lined wrought-iron pipes are also made,
but they are too expensive and hence not often used.

Drawn-lead pipe is a material possessing many merits,
and hence it is used extensively. It should be remembered, however, that some soft waters attack lead, a
sufficient amount of lead being occasionally dissolved,
particularly when the pipes are new, to cause dangerous lead poisoning in persons drinking the water from
such pipes. It is a good precaution in the case of
new pipes to allow the water to run for a while, especially if it has been standing in the pipes over night.
After they are in use for some time, lead pipes become
coated on the inside with a protective coat which seems
to prevent the dissolving of lead. Tin-lined pipes,
although more expensive, are much safer in use, but
great care must be taken in making joints in such
pipe, lest the tin be removed at the joints. Tin-lined
as well as block-tin pipes should be used as suction
pipes in wells and cisterns in preference to ordinary lead
pipes.

Plain wrought-iron pipes rust quickly, especially if
not constantly kept full of water, and water conveyed
through them is apt to cause iron stains in the washing. A further disadvantage is the frequent choking up
of the smaller sizes through rust. Pipes coated with

some kind of enamel are better and safer, provided care is taken in making the joints properly. Plain wrought-iron pipes, made rustless by the Bower-Barff process, have also been used to a limited extent for supplying water to dwellings. Wrought-iron pipes protected with a coating of zinc are used extensively, and such so-called "galvanized" pipes may be safely used, for, although water dissolves and is often found to contain salts of zinc, which are poisonous in large amounts, dilution makes them practically harmless. A more serious objection to galvanized pipes may be the fact that the zinc coating, unless applied with great care, soon wears off and ceases to protect the pipe against rust. Copper tubes, lined on the inside with tin, are occasionally used, but are expensive and troublesome to put up. In some of the Eastern States, drawn seamless brass tubes are used for hot-water pipes. Their chief advantage over lead is their greater durability, their neater appearance, and less liability to sag, although changes of temperature affect brass pipes by expansion and contraction, causing leaky joints, especially where they are under heavy pressure. Brass pipes, if used for drinking-water, should be tinned on the inside.

Plumbing Work.— Farmhouses, cottages, and suburban dwellings of moderate cost generally have but little plumbing work, especially where water is scarce, and where it has to be pumped to a distributing tank by hand labor. Where there is a more complex system of service pipes, tanks, and fixtures, there will be more or less outlay for annual repairs, besides the frequent annoyance of apparatus getting out of order, or refusing to work, or freezing up, or bursting. It is certainly much cheaper, though less convenient, to have a properly managed earth-closet and to confine the house plumbing to a kitchen sink, a force pump, a tank, and a kitchen boiler. The many advantages, however, of an indoor water-closet, as regards comfort, convenience, and health, must be conceded. A bathroom on the bedroom floor, containing a plain bath tub, is also a great sanitary convenience and an important aid to bodily cleanliness and health. Hence, it pays well to arrange for it, even where one must forego the luxury of an inside water-closet. If means are not available for putting in a system of hot and cold-water pipes, the bath tub must be filled by pails. A small slop sink or slop hopper on the upper floor for removing chamber slops is useful and facilitates the work of servants but is not needed where there is a water-closet in the house. The sink and the tub may both be arranged in one room, and this should have plenty of ventilation and direct light from a large window opening to the outer air. Even the smallest cottage should have a plain kitchen or scullery sink. Where the kitchen is large, a set of stationary laundry tubs may be arranged near the sink, because they are much more convenient than the old-fashioned portable tubs. In larger dwellings a special room is generally set aside for laundry purposes, placed next to the kitchen, or else below the kitchen, in the basement, and hot water is supplied from a kitchen boiler, heated by a waterback in the kitchen range.

Arrangement of the Bathroom. — A bathroom is always desirable, even for the small cottage or farmhouse. It should contain a wash basin, a bath tub, and possibly a water-closet, though the latter should preferably be in a

separate compartment where the floor space permits of
doing so. The arrangement of the necessary hot and
cold water pipes, waste and vent pipes should be as
plain, compact and as open — which does not necessarily
mean unsightly — as possible. A small room about
five feet by six feet will answer for a bathroom, when the
water-closet is placed elsewhere. Keep all pipes outside
of walls or partitions, have them where you can constantly see them and lay your hands on any stopcock
or other plumbing detail, if necessary. Dispense with
woodwork as much as possible and arrange every fixture,
especially the basin and the water-closet, so as to be
open to inspection and accessible to the dusting-brush
and wiping-cloth of the servants. It is important — for
the sake of economy as well as on account of plain and
straight arrangement of pipes — that the bathroom
should be as nearly as possible directly over the kitchen
so that one waste and vent pipe line may answer for the
bathroom fixtures as well as for the kitchen sink. A
little skill and foresight in planning will usually accomplish this desirable feature. The bathroom should,
however, be placed somewhat conveniently to the bedrooms and should also be accessible from the hall. It
is important that it should be warmed in winter time,
for otherwise it would be not only useless, but a source
of annoyance and expense because the plumbing would
freeze up.

Soil Pipes, Waste Pipes, and Traps. — The waste pipes
inside of the house should have joints made thoroughly
air and water tight ; they should be well flushed and
well ventilated. The house sewer inside of the dwelling,
up to a point five feet outside of the house walls, should

be of heavy iron pipe; of cast iron, if kept below the floor; of cast iron or wrought iron, if run along the cellar wall or ceiling. Always provide a sufficient number of access-holes for inspection and for removing stoppages.

The soil pipe or waste pipe should be of extra heavy cast iron, with strong hubs and well-caulked lead joints, or of asphalted or galvanized wrought iron with tight screw-joints and recessed drainage fittings. Pipes should run as straight as possible from the cellar to the roof, and be continued in full size at least two feet above the roof. The mouth at the roof should be left wide open. The following sizes of pipe should be adopted, viz.: for the soil pipe 4 inches, for the waste pipe 2 inches, for the main drain 4 or 5 inches. For branch waste pipes from fixtures use heavy lead pipe, $1\frac{1}{2}$ or 2 inches in diameter, for sinks, basins, and tubs; 4 inches for water-closets, and 3 inches for slopsinks. Make all joints between lead and iron pipe with brass ferrules or brass screw nipples.

Provide a running trap on the line of the main house sewer, placed either inside or outside of the house. Arrange a 4-inch fresh-air pipe, at the house side of the trap, and run it preferably some distance away from the house, placing the fresh-air inlet where it may be hidden from sight by shrubbery.

Each fixture should be separately trapped near its outlet by a self-cleansing and safe trap. Dispense with overflow pipes as much as possible, but if they must be used let them join the waste pipe between the fixture and the trap. Traps should be either the siphon (S or running traps), in which case siphonage must be prevented by an air pipe, or else the simpler and safer non-siphoning traps which do not require back-venting. Mechanical traps, i.e. those having in addition to the water seal a flap or ball valve, should not ordinarily be used.*

Arrangement of the Supply Pipes. — It is important to arrange all water pipes so they can be completely drained and emptied, when the supply is shut off. Pipes necessarily running on outside walls should be suitably protected against frost. In fact, in country houses it is of the greatest importance that the entire system for water distribution, comprising supply and waste pipes, cisterns, reservoirs, fixtures, and traps should be completely and perfectly protected from freezing. Even where the building is small and the work not complicated, it is well to keep for reference a plan, showing the exact size, material, and location of all water-supply pipes, stopcocks, faucets, cisterns, etc. All service pipes should be kept accessible and wherever possible in plain sight.

Water Tanks. — A tank for the storage of water is not required in cottages in which the only plumbing fixture is the kitchen sink. A suction pump is in such cases provided at the sink, and arranged so as to draw the supply either from the well or the cistern, or sometimes from both.

But where a cottage has plumbing fixtures located on the upper floor, it becomes necessary to supply these fixtures from a house tank, located in the attic or directly under the roof. Water is forced to the tank by

* For details see the author's various works on " House Drainage," " Plumbing," and " Sanitary Engineering of Buildings."

hand-labor, using a lift and force pump for this purpose, and in the case of more elaborate arrangements a gasoline or oil pumping engine may be provided, or else pumping is accomplished by means of a hot-air pumping engine.

House tanks for the storage of water should be constructed of cast-iron sectional plates bolted together and tightly jointed, or else they may be of boiler iron, with riveted joints. In both cases the rusting of the inside of the tank should be prevented by giving the tank several coats of paint. House tanks are also sometimes constructed of slate. Cheaper than all these are wooden tanks, which are lined with tinned copper. Tank linings of lead, zinc, or galvanized iron are undesirable because the water may attack and dissolve a part'of the lining. The water tank should be covered to exclude dust, flies, insects, etc. Ventilation to the roof is desirable.

From the house tank the water is supplied to the plumbing fixtures under a constant head of pressure. The house tank should always have an overflow pipe, and care should be taken not to run this pipe into any soil or drain pipe. A good way to dispose of the overflow is to run it into the roof gutter. In case such a method is not feasible, run the overflow pipe down to and over the kitchen sink, and make it answer as a tell-tale for use with the hand force pump at the sink.

More recently house tanks have been done away to a large extent by substituting for the same closed pressure tanks which can be located in the cellar or outside in the ground.

Plumbing Fixtures. — The *kitchen sink* should be a plain, or better, a galvanized cast-iron sink, with edges formed in roll-rim form. The enameled sheet iron or stamped steel sinks are not very durable, as the enamel soon chips off, and this is also, though to a lesser extent, the case with enameled cast-iron sinks. For the better class of houses a yellow or white solid earthenware or porcelain sink should be adopted.

The size of the sink should be as large as the available space permits, leaving, however, room for a drainboard, at least at one end of the sink. A kitchen sink to be serviceable, should never be less than 27 or 30 inches in length, 18 or 20 inches wide, and from 6 to 7 inches deep. On the wall directly over the sink, and the full length of the sink and drainboard, there should be a splashback, either of iron, or of earthenware, or marble, with the joints between it and the wall well filled and made solid to avoid crevices for water bugs or roaches. It is advisable to set the kitchen sink at least 2 feet 8 inches from the finished floor level, as this renders it much more convenient for use than when set lower. The sink should be supported by a pair of legs, or else be set on brackets ; the former make a stronger job, while the latter have the advantage of leaving the entire space under the sink free and unobstructed. The sink should be kept entirely open, and no wooden casing or box-like enclosure should be used. A drainboard should be provided at one or both ends of the sink ; this may be of well-seasoned ash or maple, and should be grooved and set slightly higher at the farther end so that the water and drippings from the washed dishes will run into the sink. One end of the drainboard should rest on the roll-rim of the sink, while the other should be supported by a strong bracket.

The sink should have two polished brass faucets for hot and cold water, set in the sink back at a convenient height above the sink, and the inside bottom should have a brass strainer, securely fastened to the sink with screws. The waste pipe should be heavy lead pipe, 2 inches in size, trapped by a 2-inch lead trap, with brass trap screw for cleaning purposes. It is convenient to provide at the sink a soap cup, and a rack for kitchen towels, or else a roller towel fixture.

Stationary washtubs are a labor-saving device which should be put into every house, no matter how small, as they permit the washing to be done at home and in much quicker time than where portable tubs are used. A set of two tubs is sufficient for small houses ; a more complete arrangement requires three tubs, one for washing, one for rinsing, and one for bluing.

Instead of wooden tubs, which when cheaply made are liable to leak, and when well made are almost as expensive as other kinds, and which are always unsanitary, for wood is absorbent and liable to rot, I advise using either slate or Alberene (soapstone) tubs ; some of the artificial cement stone tubs are also serviceable and cheap. Best of all are the roll-rim solid porcelain tubs, made both in white and in yellow ware ; some of the second grade of white tubs are nearly as good and serviceable as the expensive first-class tubs.

Tubs should be set about 2 feet 9 inches from the floor, and they are generally supported in iron frames. If they are placed in the kitchen, tubs with holes in the back for the faucets should be chosen, as this permits the tubs to be fitted up with covers, which in a kitchen are useful as a dresser or table. But if the laundry is in the cellar, or in a summer kitchen or shed in the rear of the house, it is better to omit the covers. Hot and cold-water supplies and faucets should be provided, also plugs and chains. The waste pipe should be from 1½ to 2 inches for each tub, and should be trapped by a trap of the same size. The tubs must always be placed where there is a good light.

The *bath tub* until recently was usually a wooden tub lined with tinned and planished copper, and provided with hot and cold-water faucets, either two single or a combination bibb, and with a brass standing overflow in place of chain and plug. A little better than copper-lined tubs are the steel-clad bath tubs, which require no other woodwork than a wooden rim, and which can be set in an open manner, on cast-iron feet, the same as the more expensive tubs of enameled iron or porcelain. Nowadays the manufacture of enameled iron tubs has been so perfected and their cost has been made so reasonable that they are used even in the cheaper class of houses. Those which are wide inside and which do not set too high on the floor should be chosen as being more convenient in use ; preference should always be given to porcelain tubs with roll-rim, and the cheaper iron tubs provided with wooden rims should be avoided.

If cost is no objection the bath tub should be provided with a plain overhead douche or spray with a hot and cold-water non-scalding mixing valve. A curtain holder may be suspended over the tub and a rubber or a white cotton duck curtain hung from it to avoid splashing over the floor when the douche is used. Such a douche or rainbath will be found of inestimable value both in summer and in winter. The floor in front of the tub, and in fact the entire bathroom floor, may be of wood covered with linoleum; in front of the tub may be placed a rubber or cork mat.

Select a plain *wash basin*, either an enameled iron oval basin, which are nowadays made almost as good as earthen wash basins, or else use the neat and inexpensive all-porcelain basins which do not require a marble slab. Support the lavatory, if small and light, on neat brackets. If a larger marble slab and a porcelain bowl attached to it are used, the basin should be supported on legs. Set the basin slab at a height of 2 feet 7 inches or 2 feet 8 inches from the floor; a lower height than this necessitates an inconvenient stooping over. Provide a back of white solid porcelain or of marble, from 6 to 12 inches high, according to the length of the slab. Avoid placing the lavatory in a corner. Provide hot and cold-water faucets and supplies, also a $1\frac{1}{2}$-inch waste pipe and trap. A lavatory is not completely furnished without at least a towel holder or shelf, and holders for soap, sponge, and tumbler, all of which fixtures are now obtainable in neat and cheap designs.

For the *water-closet* select a good earthenware flushing-rim, short-hopper closet, or better a pedestal washdown closet, or best of all a siphon jet closet, all of which should preferably have a flushing cistern. Water-closets of the "washout" pattern should be avoided. The cistern should be set from 7 to 8 feet from the floor; it should have a lever, chain and pull handle.* The water-closet should be arranged without any wood-

* Some of the new forms of low-down tank closets work satisfactorily, but the author warns against the use of any so-called "flushometer" closets having a valve instead of a tank supply, because these are rarely satisfactory in use.

work except the seat, the closet bowl standing on all sides free on the floor. A closet thus arranged answers well for use as a urinal, and also for pouring out chamber slops. The seat of the closet should be an open round or oval seat attached directly to the bowl, and hinged so that it can be turned out of the way. If the water-closet is in a separate room — which arrangement is preferable — no cover to the seat is required. In a small bathroom the cover or lid may be tolerated, as it forms a convenient seat when dressing or undressing for the bath. Provide for the closet a toilet-paper holder, those for roll-paper being in many respects preferable to those for sheet paper. The floor of the water-closet apartment should be of hard wood, or else an oilcloth or linoleum floor covering should be used. The water-closet bowl may be set up on a countersunk slate or marble platform.

There are, of course, more expensive and handsome plumbing appliances available than above described, but the fixtures mentioned, if well set and fitted up, are good enough for all practical requirements of houses of moderate cost. The author has frequently pointed out that it is very desirable to have simplicity of all apparatus and to avoid all complicated appliances. In select. ing fixtures, those made of durable material, and having durable fittings and parts, should be selected. The water-closet apparatus, in particular, should be as free as possible from all movable machinery, and other fixtures should have as few fouling surfaces and hidden channels as possible.

Sewerage. — The main house sewer outside of the building should be of strong, well-burnt glazed or vitri-

fied pipe, circular in section, laid in straight lines, or
else with curves of large radius at all changes in direction. The joints should be made with a mortar of pure
hydraulic cement. It is important that no cement
should remain on the inside of the joints where it might,
in hardening, cause obstructions. The bottom part of
each pipe joint should be made tight with particular
care. The drain pipes should be firmly laid at the
bottom of the trench, if necessary on a bed of concrete.
Grooves should always be cut out for the pipe sockets.
The drain should be laid at an average depth of three
feet below the level of the ground; all junctions of
drains should be made with Y-branches. The fall given
to the drain should, where practicable, be not less than
one-quarter inch to the foot, for pipes 4 and 5 inches
in diameter. Wherever the available fall is very
slight some simple and inexpensive flushing apparatus should be provided at the head of the line of
house sewer.

The diameter or size of house sewers for country
houses should not exceed 6 inches, and as a rule a 5-
inch pipe is sufficiently large for the purpose, while the
smaller cottages require only a 4-inch sewer. Until
recently it was the custom to use drain pipes too large
in size to be self-cleansing. To quote from an English
publication on Sewerage : "The passion for too large
pipes seems to be an almost universal one. The feeling
seems to be that it is best to make the conduit for the
sewage 'big enough anyhow,' and as a result, nearly
every drain that is laid is so much larger than is needful
that the cost of keeping it clear is often the most serious
item of expense connected with it.

"One principle is very apt to be disregarded in regulating the sizes of house sewers and of sewers in general, that is, after water has once fairly entered a smooth conduit having a fall or inclination towards its outlet, the rapidity of the flow is constantly accelerated up to a certain point, and the faster the stream runs the smaller it becomes; consequently, although the sewer may be quite full at its upper end, the increasing velocity soon reduces the size of the stream, and gives room for more water. It is found possible, in practice, to make constant additions to the volume of water flowing through a sewer by means of inlets entering at short intervals, and the aggregate area of the inlets is thus increased to very many times the area of the sewer itself. Where a proper inclination can be obtained, a pipe, eight inches in diameter, makes an ample sewer for a population of ten thousand.

"It was formerly the custom with architects and engineers to enlarge the area of any main pipe or sewer in proportion to the sectional area of each subsidiary drain delivering into it. But this is no longer done, since it has become known that additions to the stream increase its velocity, so that there is no proportionate increase of its sectional area. For example, by actual observation, it was found that the addition of eight junction drains, each three inches in diameter, to a main line of four-inch pipe, did not increase the sectional area of its flow, but made the flow only more rapid and cleansing."

Sewage Disposal. — The question of the disposal of the liquid household wastes is an all-important, but often troublesome matter. While it may be comparatively simple to dispose of the wastes from a cottage having only a kitchen sink, the difficulties are increased when country houses are provided with an abundant supply of water delivered under pressure, and utilized not only at lavatories and bath tubs, but also for the flushing of water-closets.

The subject of sewage disposal is discussed at considerable length in Part III of this book, and it is the intention to mention here merely the simpler methods. Let us assume, as is the case in nearly all isolated country houses, that there are no sewers with which connection can be made. A direct discharge of the unpurified sewage into a creek, brook, or water course, or into the ocean is rarely permissible or available. The common practice in such cases, where the soil is at all porous, is to resort to the use of leaching cesspools.

Leaching Cesspools. — The entire liquid wastes from the household are carried by means of the house drain to such a cesspool, being allowed to soak away into the soil. In a very few years the open joints of the stone side walls of the cesspool become filled with the more solid matters, with grease and scum, and the cesspool ceases to "leach." The organic matter retained in the cesspool undergoes a slow process of decomposition and creates noxious and disagreeable accumulations of gases. Usually the cesspool is unventilated, and the only exit for the gases generated in the same is through the house drain into the house pipes, and through the often defective joints and equally defective traps into the rooms of the house.

Occasionally, two cesspools are used, one for the kitchen sink wastes, the other for the waste water from the bathroom and water-closet. The conditions of the two cesspools, after they have been in use for some time, do not differ materially from each other, for the waste water from sinks, if stored in a cesspool, becomes in time as foul as any other organic liquid refuse. The arrangement of two cesspools is therefore an even greater nuisance than the one first mentioned.

Moreover, it should be borne in mind that the question is not at all simplified where a house contains no water-closet, for it is well known that a leaching or open cesspool for the waste water from sinks, basins, and bath tubs contaminates the soil around it in time just as much as if the excreta were added, the difference, if there be any, being merely one in degree.

The smaller the house lot, the greater is the danger from an open cesspool. No leaching cesspool should ever be placed nearer to a dwelling-house than one hundred feet, and it should always be put on lower ground where this exists. To locate an open cesspool close to a well which furnishes the drinking water to the household, or to a rainwater cistern, is a practice which should be forbidden by law.

Tight Cesspools. — If a cesspool or sewage tank must be used it should be built thoroughly water tight. It should be of moderate dimensions, preferably circular in shape, built with hard-burnt brick, laid in hydraulic cement mortar and the tank should be well rendered inside and outside with pure Portland cement. The tank should be arched over and covered with a tight iron cover. The cesspool must be emptied, cleaned, and disinfected at frequent intervals, and it should, if possible, be well ventilated. The cesspool should preferably not be located in a direction from the house from where the prevailing winds blow.

An important modification and improvement upon this plan consists in building the cesspool in two chambers, the first one being intended as a retaining chamber for grease, or for the solid organic matter; from this intercepting chamber the liquid overflows into the second chamber.

Sewage Irrigation. — It is much to be preferred to arrange for a distribution of the liquid contents of a cesspool or sewage tank on or near the surface of the ground. In this way sewage may be used to sprinkle and irrigate a lawn, a kitchen garden, a group of shrubs, or a vine trellis. The intercepted solids should be removed at frequent intervals and may be dug as fertilizers into the ground. If this arrangement is adopted the above described construction of the sewage tank in two chambers should be followed, the smaller of the two chambers being for the solids and the larger for the liquids. The overflow pipe connecting both chambers should dip several feet below the water level of the intercepting tank, so as to avoid carrying scum or grease with the water. The liquid manure may be pumped out by means of a small cesspool pump, set over the top of the liquid cesspool chamber.

Earth Closets. — The question of the disposal of the wastes from the household is, to a certain extent, simplified, and the liquid manure is more easily removed and taken care of, where the cottage contains no water-closets. But, in that case, it has been usual to substitute for a water-closet a privy vault located either close to the house, and thus constituting in itself a formidable nuisance, or placed at the remotest corner of the lot, in which case it is rendered difficult of access in stormy weather and in winter time generally. In any case, the privy rivals with the leaching cesspool in nastiness and danger to health. It pollutes the soil, taints the water in the well, and contaminates the air of the whole neighborhood. Privy vaults should always receive unqualified condemnation. In the various appliances known as earth or ash closets, we find a better and cleaner substitute for the privy.

I do not feel justified in recommending the use of an earth closet inside of a cottage, except for the use of invalids. It is, however, not very difficult to arrange it so as to be quite near the rear part of the house, and to make it accessible from the house by a not too conspicuous, well-covered, shady, dry, and sheltered walk. The shed, in which the earth closet is placed, should be well built, strong and tight, and preferably plastered, so as not to be too cold in winter storms, but it should also be sufficiently ventilated. Expensive forms of earth closets, with mechanical apparatus for throwing a fixed quantity of earth after each use, are sold and generally give satisfaction when used intelligently, but in the case of cottages of low cost a plainer form of this closet answers the purpose fairly well. The earth manure can be advantageously used in the kitchen garden or else it may be disposed of to neighboring farmers.

Regarding earth closets, I quote the following from the annual report for 1892 of the State Board of Health of Maine:

"All that is needed is a common closet, a supply of dry
earth, a water-tight receptacle beneath, and a convenient way of disposing of its contents at quite frequent
intervals. The receptacle should be wholly above the
surface of the ground, and may consist of a metallic-
lined box, a half of a kerosene barrel with handles on
it for removal, or still better, a galvanized iron pail.

"The receptacle may be removed through a door in the
back of the closet, or in front of the seat, or by having
the seat hinged and made to open backward it may
be removed in that way.

"The earth should be common garden or field loam finely
pulverized. Road dust does well, but sand is not suitable. Coal ashes are also good. Whichever of these is
used should be dry and screened through a sieve with
about quarter-inch meshes. The dry earth may be kept
in a box or bin so arranged, where it can be, that it may
be filled from the outside of the closet, or it is quite
convenient to have one-half of the seat hinged, and
beneath it the small compartment to hold the supply
of earth. In this box or bin holding the earth there may
be a small tin scoop which may be employed in sprinkling the earth, a pint or more each time the closet is
used. The main thing is to use enough of the earth to
completely absorb all liquids, and this requirement, of
course, precludes the throwing of slops into the closet.

"Arrangements could easily be made with gardeners or
with farmers for the daily removal of the contents of
these receptacles for fertilizing purposes."

Slopwater Disposal by Sub-Surface Irrigation. — The disposal of the slopwater, consisting of the chamber slops
and of the kitchen water, may be effected where there
are grounds about the house, which slope somewhat
away from it, by the method usually known as the subsurface irrigation system. This consists in placing a

series of common 2 or 3-inch unglazed and porous drain
tiles in parallel lines in trenches, dug to a depth of
from 10 to 12 inches below the surface. These absorption drain tiles are laid with open butt joints with a
space of at least one-quarter of an inch between the
tiles. Thus the sewage is distributed intermittently
through a network of pipes into the upper layers of
the soil, where it is acted upon by the nitrifying organisms contained in the soil, while the vegetation assists
in the process. The liquid from which the grosser
organic impurities have been removed is further purified
by filtering through the soil.

A sufficient number of tile lines, of moderate length,
and not exceeding one hundred feet each, must be provided to obtain the required capacity in proportion to the
volume of daily sewage. It is absolutely required that
the flow in the absorption drain tiles should be intermittent, for if a continuous dribble of sewage occurs,
the tiles are sure to choke up in a short time, and the
soil will become surcharged and swampy in single spots.
Therefore, it is advisable to use a sewage flush tank
in connection with the system. The sewage from the
house accumulates in the tank and when this becomes
filled it discharges automatically, at certain intervals.
Where the quantity of sewage is small, the discharge
of the tank can be economically effected by means of
a gate valve placed on the outlet drain leading from the
bottom of the tank to the irrigation field. In this case,
hand labor is required to operate the system. The
alternative is to use one of the many forms of automatic
sewage siphons. Such a system answers admirably for
the inoffensive disposal of the sewage from isolated

country houses. It is equally practical when water-closets are fitted up in the house, but in this case the solids should be intercepted in a small receiving reservoir or chamber, which requires frequent cleaning, otherwise the distributing tiles may speedily choke and create a nuisance in the disposal field by ceasing to perform their work properly.

Sewage distribution on the surface, while somewhat simpler, requires a larger area of ground, and cannot be carried out in the immediate neighborhood of houses. It is also apt to be somewhat more troublesome in winter time.

Other methods of sewage disposal for country houses comprise septic tanks, contact beds, and trickling filters. Reference to these more recent biological disposal methods is made in Part III.

Garbage Disposal. — In small country houses, the problem of the removal and disposal of the kitchen and house refuse is simple, as a rule, for much of the kitchen offal can be readily disposed of to farmers, while parts of it may be utilized on the home grounds, where a few domestic animals are kept, or else they may be thrown in shallow layers into a trench, dug for this purpose, and covered up with some lime and earth.

If the house has a large set range, in which either wood or coal is burned, a large part of the kitchen garbage may be disposed of by drying and carbonizing it in special household garbage destructors, after which it may be thrown into the kitchen fire.

Sweepings, old dust rags, and other dry refuse may be best disposed of by direct burning in the range.

Household Garbage. — It is inadvisable to burn fresh kitchen garbage in a range, for it contains a large amount of moisture and the process would destroy the linings and fittings very quickly and would give off offensive smoke and smell in the kitchen. A much better way is to dry and desiccate the garbage, to carbonate it by heat, and then to burn it up in the fire. A practical device for accomplishing this purpose, the construction of which is based on sound principles, is the *household garbage carbonizer*, which consists essentially of two parts, namely: first, a horizontal cylinder or drum a little larger in diameter than the size of the smoke pipe, which cylinder is inserted permanently in the smoke pipe between the range and the chimney flue; second, a removable front piece in disk shape to which is attached an inside perforated basket or tray. When this front end is inserted into the drum the area of the space around the basket is somewhat larger than the area of the smoke pipe and therefore the draft is not obstructed.

The use of this device is simplicity itself. The basket with the fresh garbage is inserted into the drum. The fire gases and smoke from the kitchen range pass around and through the basket, drive off the moisture in the garbage and slowly carbonize the same, *i.e.*, they reduce it to charcoal. The resulting product is very serviceable as a kitchen fuel. No fat should be put in the basket, as there would be danger of its catching fire.

The entire carbonizing process is accomplished without any objectionable odors, and the carbonizer does not interfere in the slightest with the regular use of the range. No extra fuel and but very little attention is required. The device can be applied to any form of coal or wood.

burning stove or range, and it is manufactured in various sizes and styles. Experience shows that the servants soon learn to appreciate the advantage of the device because it saves them labor in the kitchen. The fact that this appliance can be successfully applied to all kitchen wastes (except fat) and that the method is convenient, clean, and not costly should render its success assured. If every household were provided with a carbonizer, the work of the scavenger would be reduced to a minimum, and but little, if any, household garbage would have to be carted through the streets.

The garbage carbonizer accomplishes three things in the following order, namely: first, it dries the garbage by driving out the moisture; second, it changes the dry garbage into charcoal; and third, it burns the latter in the kitchen fire. It not only destroys the household wastes without nuisance, but it transforms them into useful fuel. It is a practical, simple, convenient, effective, inexpensive, and easily applied device, which does away with the nuisance of the garbage can. To the kitchen servant the device saves a good many steps daily by doing away with the necessity of a garbage can. It prevents the accumulation of putrefying matter in dwellings, and in cities it renders unnecessary the unsightly garbage carts passing through our streets. The very simplicity of the device should render it attractive to all housekeepers.

The device described was put on the market some years ago and proved very successful in use. It is much to be regretted that the appliance is no longer manufactured. It would seem as if a useful device such as the one described would constitute a profitable one to manfacture.

A similar so-called domestic garbage burner is now made in Kalamazoo, Mich., which also can be attached to all kinds of ranges and stoves including gas and gasolene stoves. It has a tilting hopper into which the kitchen garbage and sweepings and litter are put. As soon as the hopper is closed the waste heat from the fire passes through and around the hopper and reduces its contents to carbon and ashes. In this way potato parings, egg-shells, bones, bits of meat, rags, melon rinds, and other offal can be readily destroyed. This device helps much in the same way as the one first described to solve the vexed question of garbage disposal. Destruction by fire of all household garbage is certainly vastly better than its temporary storage on the premises, which invariably attracts flies, ants, vermin, and other insects.

Another household apparatus for the disposal of garbage is the Victor cremator, made in Columbus, Ohio. This is operated by means of gas, and consists of a cylinder fourteen inches in diameter and twenty-four inches in height, which is lined on the inside with pure asbestos to retain the heat. It can be placed according to convenience, either in the kitchen or in the cellar, and requires a gas supply and a chimney connection. The apparatus consumes about twenty-five cubic feet of gas in half an hour and the average daily amount of garbage in a household is said to be consumed in from thirty to forty-five minutes. Properly used such a crematory outlasts several garbage cans, which wear out rapidly from the rough handling and the exposure to the weather, and are generally unsanitary, unsatisfactory, and offensive. After operating the device nothing remains but a heap of ashes, which can be readily disposed of. This crematory appears to offer a practical solution for the economical and sanitary disposal, not only of kitchen garbage, but also of litter, sweepings, and other combustible household wastes.

Garbage which cannot be disposed of in the manner indicated should be stored temporarily in impermeable, non-corrosive, covered cans or pails, and should then be removed as often as required.

Hints on the Care of the House. — Home sanitation involves the constant maintenance of cleanliness in all parts of the house. Those which require the largest amount of care and attention and which must be scrupulously looked after to remain perfectly sanitary, are the places where the food is prepared, or where food supplies are stored, such as the kitchen, the pantry, the cellar, the storeroom, and the refrigerator. Plumbing work in bathrooms, kitchen, pantry, and laundry should also be constantly looked after and kept bright, neat, and clean.

Both in summer and in winter the house should be flooded daily with air, light, and some sunshine. To keep rooms shut up and dark for days or weeks at a time cannot conduce to sanitary conditions. The heated period of summer is the time when the greatest care, perhaps, should be exercised, for at this time conditions are apt to arise in and about a house which may seriously menace the health of the occupants.

Everything possible should be done to keep flies out of a house by providing window screens and self-closing screen doors. For the kitchen, the pantry, and the dining-room, screens are absolutely essential, for I have pointed out elsewhere that the germs of preventable disease may be transmitted to the food by means of flies.

In and about a country house all pools of stagnant water should be done away with, and rain-water cisterns should be tightly covered, or provided with fine netting. Such places are known to harbor and breed mosquitoes, which as I have already mentioned are able to transmit malaria, while some species, fortunately found only in the Southern States and in tropical countries, are the carriers of yellow fever.

A prudent house owner should make a weekly inspection of his cellar, noting particularly the cleanliness of the walls and floors, and attending to the removal of any matter which is perishable and liable to decay. Once a year the walls of the cellar should receive one or two coats of whitewash and the floor should be swept at least once a week, preferably with a cloth-covered broom which has been moistened by means of some disinfecting solution. Instead of this it may answer to sprinkle from time to time some liquid disinfectant over the floor. If there is a sink in the cellar or any cellar and area floor drains, these should be carefully looked after and a disinfecting liquid poured into them. Deep cellar areas are not infrequently the breeding places for mosquitoes, and to guard against them it is advisable to pour from time to time some kerosene oil into the area drains.

In the management of the cellar of a house one should always remember that dirt or filth of any kind may breed disease, and that the air of the cellar passes not only through the cracks in the ceiling upward into the house, but that it is also liable to be drawn in at the furnace cold air box if this is not perfectly tight in the joints and thus be sent up in a heated condition to enter the apartments of the upper floors through the registers.

The air of the cellar should be purified from time to time by keeping lime or charcoal exposed in a shallow vessel. The cellar should be well aired by opening the windows daily, and this on days when the outside air is not only very warm but also laden with moisture, should be done only at night to prevent the condensation of the humidity on the colder cellar walls, which causes mildew and dampness of floors and walls.

If there is an outdoor earth closet, this should receive care and attention at least once a week. No liquid wastes or slops of any kind should be disposed of by throwing them into the earth closet, neither should they be dumped on the ground near the house, nor near the well or the cistern.

If there is a well, it should be at all times guarded against surface pollution. The soil, the water, and the air around a country house should be kept pure and undefiled.

Pails or other vessels used for the storage of garbage should be scrubbed once a week with hot water and soapsuds, or with water and ammonia or soda, and afterwards they should be aired outside of the house. The occasional use of a disinfecting solution or of carbolic acid in powder form is advised.

It is of the greatest importance that proper care be given to the refrigerator and to the place where it stands. The refrigerator should always be set on castors and be portable or removable; built-in ice boxes or refrigerators are not to be recommended. Once a week, at least, the interior of the refrigerator should be thoroughly cleaned, the shelves taken out and scrubbed with hot suds and the whole exposed to the air and the sun

All corners, in which fragments of food or spillings are liable to accumulate, and the drain or waste pipes which carry away the water from the melting ice, should be particularly looked after. A little caustic potash dissolved in very hot water should be poured through the waste pipe to keep it free from obstructions and to prevent its ultimate stoppage. The tray or pan in the floor under the refrigerator collects in a short time a good deal of slime from the ice and dust from the air, and, therefore, should be well cleaned and cared for. Milk and other articles of food spoil very rapidly and become unfit for use where a refrigerator is not well taken care of and kept in a sanitary condition. A well-kept refrigerator should be entirely odorless and it should be remembered that the low temperature of the ice box prevents any odor from becoming very pronounced, but as soon as the ice is all melted away the peculiar smell from an ill-kept refrigerator readily announces the fact that decomposition goes on in the interior of the box, and the warning thus given should never be neglected.

Dusting and Sweeping.— In large cities an enormous amount of dust is constantly generated, disseminated, and introduced into the houses, but even in the country the localities are exceptional where dust does not come in from outdoors to settle on the carpets, furniture, books, upholstery and curtains.

Country houses located close to a highway are notable sufferers in recent years, owing to the rapidly increasing number of automobiles, and this is particularly the case where the speed of such vehicles is not limited to a moderate amount, wherever they pass near dwellings.

Indoor dust has been well called the bane of the tidy house-keeper, while outdoor dust may become an extremely annoying factor to the pedestrian. That dust is a serious nuisance cannot well be disputed, but what is of much more importance from a sanitary point of view is that dust may become dangerous as a factor in the production of disease, particularly of the organs of breathing. Street dust may also become a source of danger if it gathers and settles on the food stuff which is exposed in front of their shops by grocers and others.

We may distinguish between street dust, industrial, and domestic dust. Much of the outdoor dust is due to the bad practice of burning soft coal, but all floating dust also carries a large number of bacteria and it may carry germs of disease.

Much can be done to reduce the dust nuisance outdoors by adopting smooth and easily cleaned paving materials, by using tightly covered ash and garbage carts, by not permitting the dry sweeping of streets, nor the sweeping of the dust from houses, workshops, or stores into the streets.

Indoor dust can to some extent be prevented by abolishing carpets and using instead loose rugs or mats; by adopting building and floor construction which do away, at least to a certain extent, with accumulations of dust underneath the floor boards; by adopting hardwood floors and using few light hangings instead of heavy curtains and fluffy upholstery.

In many households antiquated methods of cleaning still prevail. The use of the feather duster should be either abolished entirely, or else it should be permitted only when all the windows can be opened wide. It is far preferable to make use of a moistened dust cloth. In the same way the ordinary method of floor sweeping should be condemned and the wet mopping up of floors should be recommended. Carpets should be sprinkled with moist tea leaves, or damp sawdust, or coarse salt before they are swept. The improved carpet sweepers are better than ordinary brooms, and for well-to-do households the new methods of vacuum cleaning, of which there are several systems, portable as well as stationary, are worthy of being looked into.

In winter time when the heating apparatus is in operation much dust enters our houses through the cold air box, and a great improvement can be effected by adopting cheese-cloth screens for filtering the air supply.

The Care of Plumbing Work. — A few words should be said about the proper care of plumbing work. The term, as used here, should be understood to signify the precautions which it is necessary to take to keep plumbing work in good and efficient working order, and in a sanitary condition. The simple matter of the mere cleaning up and polishing of the brass and nickel fittings and trimmings of plumbing appliances will not be referred to in the following hints.

In order to maintain a plumbing and house drainage system in a good condition it is necessary, above all, that stoppages in waste pipes, in traps, or in the house sewer and its various branches be avoided.

Where the drain and waste pipes are kept exposed in the cellar, and are provided with brass clean-out screws, it is a good plan to have these opened up once a year, and to run a swab through the drain or waste pipe to remove any accumulation of solids. Care should be taken to replace the cleaning screws and to close them perfectly air tight.

All traps, including the main house trap where such is used, should be cleaned out at regular intervals, and this precaution is very necessary in the case of traps for kitchen or pantry sinks and for laundry tubs, the former being liable to accumulate grease or coffee grounds, while the latter are apt to catch and hold back much lint from the wash. In this way bits of rags or strings, hair combings, lint from towels or napkins, pieces of absorbent cotton, burnt matches, etc., are removed, which if left may either clog the trap and the waste, or lead to the unsealing of the trap by capillary attraction. Matches, in particular, should not be permitted to be thrown into any plumbing fixture, as they often lodge in the trap and thus lead to accumulations of solid matters.

The strainers of fixtures require a weekly cleaning and looking after, for they also collect lint and hair, which are apt to clog the outlets of the fixtures.

In the case of kitchen sinks, the most frequent source of trouble is grease, and next to that tea leaves and coffee grounds. Grease should never be permitted to run to waste in a sink; it should be scraped from the pots, pans, and table dishes and then collected in a special receptacle for removal or disposal. In the same way it is advisable to catch tea leaves or coffee grounds by means of a suitable sink strainer and sieve, and to throw them into the garbage can. No solid food scraps or waste bits of meat should be disposed of through the kitchen sink.

All traps should be frequently flushed with plenty of

water, preferably hot water. After the use of a wash basin, bath tub, or sink, it is a good practice to let some clean water run through the fixture, and thus to leave *clean* water standing in the trap until the next use of the fixture. The occasional use of caustic potash dissolved in boiling water is recommended as being an excellent and cheap method of keeping pipes from stopping up. Occasionally, some liquid disinfecting solution should be thrown into the traps.

Iron sinks are kept clean with kerosene, or with soap and water. Porcelain sinks and enameled iron sinks should never be cleaned with sapolio or similar gritty cleansing soaps, but some of the special preparations, which are now readily available and widely advertised, should be used.

In the case of bath tubs and wash basins, the inaccessible overflow pipes, and the hidden or secret waste valve fittings are the most usual sources of bad odors, which latter are due to accumulations of decaying soap. The secret waste valves or plugs, which are arranged to be lifted out, should be washed every week with ammonia and warm water. The overflow holes of the basins or tubs should be cleaned as far as this is practicable ; the later forms of basins have detachable metal overflow strainers, and in these the overflow may be kept tolerably clean and inoffensive.

Water-closets should be washed and flushed once a week with boiling water and concentrated lye. Once in a while, a solution of copperas, or of carbolic acid, may be poured into the closet bowl and trap to disinfect them. The wooden seat and the marble floor slab on which the modern closet stands require particular attention.

Solid porcelain and enameled iron bath tubs are kept clean and bright by the use of special cleansing powders which do not scratch the enamel surfaces.

In cold weather, care should be exercised to prevent the freezing of pipes, traps, or fixtures. Householders realize the importance of this matter but builders do not always give it sufficient attention. The result of the carelessness of the builder or of the plumber is annoyance, trouble, and expense to the householder. Where plumbing is imperfect in this respect, all exposed pipes must be shut off *and emptied* in cold weather, particularly at night, when no water is drawn for use. All accessible pipes in exposed positions should be well wrapped with wool felt, or covered with magnesia-asbestos coverings.

Particular care should be taken with plumbing work in those country houses which are left closed for the winter. All pipes, traps, and fixtures should be completely drained and left empty. Equally necessary is the protection of plumbing work in houses left vacant during the summer. In this case, it is required to shut out the sewer air by providing protection by other means than those afforded by the water seal of the common forms of traps. Olive oil and glycerine are largely used to fill the traps in preference to kerosene oil which is too quickly volatilized, and hence affords no security for a greater period than two or three weeks.

House Disinfection. — While, as a rule, house disinfection is only practiced after the occurrence of a case of infectious disease, it would seem to me that it would be advantageous to practice disinfection at other times. It would, for instance, seem to be advisable not to move

into a house just vacated by other tenants, whose habits
of cleanliness and conditions of health may not be known
to the new tenants, without first applying some disinfectant, if not to all the rooms, at least to the bedrooms.

A convenient way of disinfecting a room is by means
of the Scheering disinfecting and deodorizing lamp,
in which formalin pastilles are evaporated.   The advantage of this disinfectant is that the fumes are neither
poisonous nor destructive to furniture, clothing, or books.
Disinfection by means of sulphur burned in candle form
is also quite frequently practiced.   Sometimes disinfection
is applied in a house in order to rid it of some animal
pests, like cockroaches, fleas or bedbugs, and a word of
caution seems necessary where such disinfection is
carried out by means of the fumes of hydro-cyanide of
potassium, because the fumes are a deadly poison.

Stationary ice boxes, as well as portable refrigerators,
should be washed out from time to time with a formalin
solution, or disinfected by means of the lamp and the
formalin pastilles.

It is furthermore an excellent practice before the furnace is lit to use in the evaporating pan some liquid formalin diluted with water, or else the Sanitas disinfecting
liquid.

NOTE.— Much practical information on the subjects discussed in this
chapter may be derived from a perusal of a most excellent little up-to-
date pamphlet, entitled "Modern Conveniences for the Farm House"
The author is Miss E. T. Wilson, C. E., and the pamphlet was prepared
in 1906 for the United States Department of Agriculture, and is known
as, "Farmers' Bulletin No. 270."

II.
WATER SUPPLY

THE WATER SUPPLY OF COUNTRY HOUSES

In the case of buildings located in the country the engineer engaged in the problem of water supply undertakes not only the instalment of the *inside* water supply, but he also plans and carries out the entire *outside* water supply system.

Engineering Advice. — For country homes located beyond the limits of city water works, the need of a pure, reliable, and ample water supply should always be one of the chief considerations. In order to advise intelligently on the methods of obtaining it and to prepare the required plans, a great many points have to be considered. The advantages and disadvantages of the different schemes which offer a possible or a suitable solution of the problem must be carefully weighed before making a decision. Rather than make a decision for themselves and incur the risk of failure, owners of country mansions and estates should seek the disinterested advice of a competent hydraulic engineer.

Points of Importance in Studying Problems of Water Supply. — The order in which the different points of importance are taken up is usually the following: — The available *sources of supply*, be they springs, wells, brooks, rivers, lakes, fresh water ponds, surface waters from dammed-up water sheds, or collected rainwater, must be examined. The *location* and *elevation* of the proposed source of supply with respect to the site of the buildings to be supplied usually determine the question whether a *gravity supply* or a *pumping system* must be adopted.

In determining upon a source of supply, it is well to bear several general facts in mind: first, that the nearer the source is to the buildings, the smaller will be the cost of the water supply system, but the greater becomes the risk that the water may be, or may become, unfit or dangerous to use; second, that surface sources, such as springs, dug wells, shallow driven wells and cisterns, yield, as a rule, a supply which is limited in volume; third, that where larger volumes of pure supplies are required, deep or Artesian wells should be provided, which are not, as a rule, liable to pollution from surface impurities.

The *quality* of the water proposed for use is of the foremost importance because impure water, containing disease germs, is a known vehicle of a large amount of preventable disease. The determination of the quality of the supply involves chemical and bacteriological analyses, as well as examinations of the water sources, of the surroundings, or of the watershed. As a rule, mountain springs, brooks and lakes in uninhabited regions, and deep wells yield pure and wholesome water; water from shallow wells and from rivers or ponds must be considered dangerous and unsafe unless its purity is established beyond the shadow of a doubt; cistern water and surface water from cultivated farm lands should be regarded with suspicion.

Means for the prevention of the contamination of the source of supply must in many cases be provided; and on the other hand, where a water is already slightly polluted measures or methods for its purification must be designed and planned.

The *quantity* of water available from the source selected must be compared with the quantity estimated as the probable future consumption. This is quite important, because many of the available sources of water supply in rural districts are limited or scanty in volume.

It is likewise necessary to determine, at the outset, the *water pressure* required for domestic use and also for fire protection. If the water must be pumped, various systems may be adopted, such as pumping to a reservoir, to an elevated tank, to a standpipe, to house tanks, or else water may be pumped directly into the supply mains, or finally it may be pumped into pressure tanks.

This subject naturally leads to the consideration of the next matter of importance, which is the determination or selection of the *pumping plant*, and of the power or *motive force* to be used *for lifting the water*. This, in the case of smaller farm buildings or cottages, may be either hand or animal power; in the case of larger country houses, the motive power may be obtained from gas, hot air, gasoline, oil, electricity, the force of the wind or the power of falling water; in the case of very large or extensive buildings, or groups of buildings, the motive force may be either steam or electric power.

The water after being pumped from the source of supply must be *stored in reservoirs or in tanks*. Here again, several methods are available, such as the construction of a stone or earth reservoir, the erection of wooden or iron tanks on wooden, iron, or masonry supports; the building of a tall standpipe, which may be either open or enclosed, and finally the use of pressure tanks, located either in the cellar or outside of the building, in the ground.

After being pumped and stored, the water must be

brought to the buildings where it is to be used. This involves the consideration of both the inside and the outside water supply *distribution systems*, and the arrangement in detail of the water piping. Of the greatest importance in the case of country houses or institutions, located far away from the protection which a good city fire department affords, is the *provision for indoor and outdoor fire protection appliances*, for which an ample volume of water must be instantly available.

The various points outlined in the foregoing will now be taken up more in detail.

Springs. — Springs are natural outlets at which underground water flows out at the surface. In some parts of

FIG. 1. — A SPRING AS A SOURCE OF WATER SUPPLY.

the country springs yielding a pure water are numerous and are considered a valuable source of supply (see Figs. 1 and 2). Spring water is usually palatable, wholesome, pure, and free from organic impurities, owing to the natural filtration going on while it passes through the subterranean strata and before it crops out, to form a spring. It is, of course, best and most convenient if springs are found

FIG. 2. — A SPRING ISSUING FROM THE ROCK.

located at an elevation considerably higher than the house and grounds to be supplied, because in that case the water may flow by gravity. If a spring emerges from the soil or rock at a lower level than the house, some form of pumping machinery is always required. But if its volume is larger, its yearly flow more uniform, and its character better than that of a spring at a higher elevation, it may be advisable to prefer a pumping scheme.

In contemplating a water supply obtained by tapping a spring, it is of the greatest importance to give attention to the variation in the yield of the spring, which occurs almost regularly at the different seasons of the year. It often happens that, when a spring is selected which was examined perhaps during one of the wet months of the year, generally the spring months, it is found, later on, that its flow becomes so reduced, during a dry summer or in the early autumn, as to be quite insufficient for the requirements of the building. It is obviously much safer, in all cases, to defer the examination or the gauging of the spring until after dry weather has set in, and to select a spring only in case it yields, even at such times of drought, an excess of water over the maximum supply required. In general, the yield of a spring depends upon the rainfall of the region, and upon the area of permeable strata which supply the spring. The more distant from the spring this area is located, the more regular its flow is apt to be; but it is rarely possible to determine, even after careful inspections, where the area supplying it is located, or what its size may be.

For the protection of the spring against contamination it is necessary to wall it in; it is still better to construct a small storage basin in which the night flow of the spring can be stored and which holds a reserve of water. This basin should have a tight cover, or a house should be built over the spring basin, to exclude surface impurities, dust, and animals, and to protect the spring against careless or malicious contamination. The pipe leading from the spring basin to the house should have a strainer, to keep back the larger substances in suspension, such as leaves, sticks, or other débris.

"The freedom of many spring waters from deleterious mineral ingredients is well known. Many spring waters undergo a natural filtering process through the soil and are therefore clear and limpid, soft and excellently well adapted to the needs of the household.

"There are, however, many springs that are liable to

contamination from the wastes from human habitations, and this is particularly true of those copious flows which issue from the base of alluvial terraces along the larger rivers. These terraces are in many cases covered with less than twenty feet of loam, sandy and gravelly at the bottom, resting on a bed of clay or marl. Where such a terrace is dotted with dwellings the spring flowing from the porous bed may receive, after slight and altogether in-sufficient filtration, disease-laden filth, and the freer the passage of water, the more copious the flow, the greater the danger. No matter how clear and sparkling such water may appear or how alluring its coolness on a sultry summer day, it may contain the germs of typhoid fever or other water-borne diseases and its apparent purity may be a ghastly sham." (From Bulletin United States Geological Survey.)

Wells. — Wells are used in the country probably more than any other source of supply. Water from wells is really rain water which has percolated through the soil down to a water-bearing stratum; it is rain water purified in some cases by a natural filtration through percolation in the soil, but in other cases changed in character by reason of having taken up soluble mineral constituents from the geologic strata through which it flows. Well water may be regarded as underground water the same as spring water, but in the case of a spring, as we have seen, the water has found or forced a natural outlet whereas wells constitute artificial outlets, for they have to be sunk, dug, driven, or drilled to the water-bearing stratum as the case may be, and moreover the water from wells must be lifted by pumping except in the case of the flowing wells.

Although it is usual to distinguish between shallow and deep wells, there is no sharp demarcation between the two, and the designations "shallow" and "deep" really refer more to the nature of the underground strata into which a well is driven than to the depth of the well. Wells which are sunk to a depth of say from 15 to 30 feet to water flowing in a superficial layer of gravel or sand, which in turn rests upon an impervious stratum, are considered to be shallow wells; on the other hand, deep wells are those which go through an impervious stratum or through rock, in order to tap a different water table at a greater depth.

Wells in Rock. — Regarding wells driven through rock, the following information given by the United States Geological Survey is of interest:

"The very general belief that wells sunk in granite will get no water appears not unreasonable when we consider that granite is the hardest of rocks and that its surface outcrops are, as a rule, so free from pores or crevices through which water might circulate that the expectation of finding water by drilling would seem absurd. Within the last few years, however, the many successful wells drilled in crystalline rocks have effectively proved the erroneous character of the old opinion.

"The crystalline rocks, such as granites, gneisses, schists, etc., like the sedimentary shales, limestones, and sandstones, carry water in their pores — the microscopic spaces between the grains of solid mineral matter; but while the sedimentary rocks may absorb several per cent of their volume in water — sandstones about 15 per cent, limestones 5 per cent, and shales 4 per cent — granites and other crystalline rocks rarely absorb water to more than one-half of 1 per cent of their volume, and the water in such rocks moves through the pores so slowly that it can never escape fast enough to be of value in wells. Fortunately, however, the crystalline rocks are traversed in various directions by many joints and crevices. An investigation of these joints shows two principal systems, one of which is nearly vertical and the other horizontal. The vertical joints, which may be hundreds of feet in length, while not at all regularly spaced, are usually 10 to 20 feet apart, trend in all directions, and may be inclined to the vertical at any angle up to 30 degrees. The distance between the horizontal or sheet joints, which approximately parallel the surface, varies from a few inches near the surface to many feet at a depth of several hundred feet below.

"The nearly vertical joints serve as channels for the admission of water from the surface, while the sheet joints form reservoirs for its storage. As most of the joints are rather narrow, the amount of contained water is likely to be moderate, and the yield of wells in granite is seldom more than 10 gallons per minute, though some exceptional wells, pumped by steam, have yielded as much as 30 gallons a minute. Out of 72 successful wells in northern Maine only two yielded more than 50 gallons per minute.

"The extreme irregularity of the joint systems makes the success of any well in granite a matter of chance. Of two wells drilled within 50 feet of each other one may be a failure, the other a marked success. Records collected in southern Maine indicate that about 87 per cent of the wells drilled in granite supply water enough for ordinary domestic uses.

"The depth to which drilling should be carried in granite, as indicated by the investigations made both in Maine and Connecticut, has a maximum limit of about 200 feet, below which the chances of success diminish rapidly. Out of 47 wells reporting the principal water horizon, 17 or more than one-third, found it within 50 feet of the surface; 16 more, or over one-half, of the remainder, reached it within 100 feet of the surface; 7 or exactly half of the wells more than 100 feet deep, tapped their principal supply between 150 and 200 feet; and only 4 wells, or 50 per cent of the entire number of recorded wells in granite over 200 feet deep, obtained it from lower depths.

"Granite water, where not contaminated by surface drainage, is excellent for drinking and is probably satisfactory for all other uses ordinarily made of water.

"As to the cost, which is necessarily an important factor in the sinking of wells, the price for drilling 6-inch wells in granite on the coast of Maine ranges from four to six dollars

per foot, the higher figure being the more common. The cost
of drilling granite wells and blasting open wells in the same
rock, is in some sections of Maine the same for either type,
six dollars per foot. Its freedom from the great danger of
pollution, common to all open wells, makes the drilled well
in every case preferable.

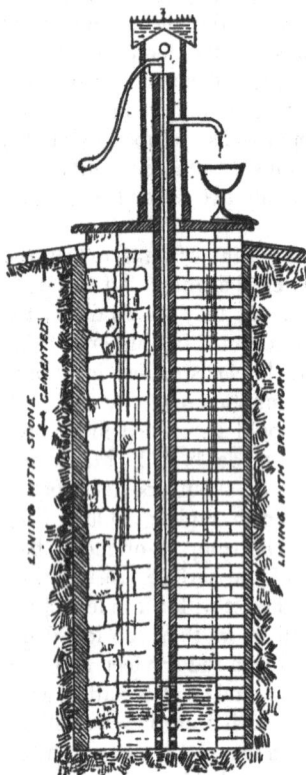

FIG. 3. — A DUG SHALLOW
WELL.

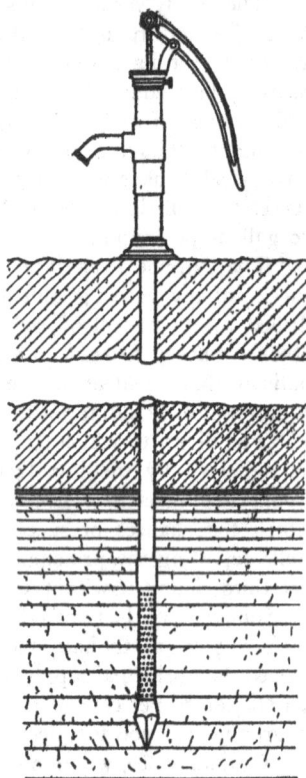

FIG. 4. — A SHALLOW DRIVEN
OR TUBE WELL.

Shallow Wells. — Shallow wells should always be looked
upon with suspicion because they are liable to become, or

to be, polluted, particularly in populous districts, but not to a lesser degree on the farm, if they are located close to outhouses, cesspools, stables for horses, or cow barns. The water from shallow wells is rain water which has percolated through the permeable surface strata and which lies over or on some underlying impervious stratum.

Shallow wells are constructed either by digging a hole of sufficient diameter, and lining it after completion with stone or brick walls — the so-called "steining" of wells — (see Fig. 3), or else they consist of small wrought iron tubes $1\frac{1}{2}$ to 2 inches in diameter driven into the ground to the depth of the underground supply (see Fig. 4). It is worth mentioning that the diameter of a well has not so

FIG. 5. — A BUCKET OR DRAW WELL.

great an influence on the yield of water as commonly thought, and very large supplies may be obtained from wells only 4, 6 or 8 inches in diameter.

In general, bucket or dip wells are not as good as pump wells, because they are more readily polluted by surface impurities, or by the pail, bucket or other form of vessel dipped into them to raise the water to the surface (see Fig. 5). All dug wells should be periodically cleaned.

Contamination of Wells. — It is of the greatest importance, in locating a shallow dug or driven well, to examine first into the direction of the flow of the underground water, in cases where the drainage of the house depends upon the use of open cesspools, a method which, as will be pointed out in Part III, is never to be recommended. Along the southern shore of Long Island, many wells on farm properties may be found located "above" the cesspool, the contention apparently being that any soakage from the latter into the underground water course will not contaminate the well, as the flow will be away from the well. In this locality, one frequently encounters the assertion that those wells which have a source of contamination *above* them are dangerous to health, whereas those which have cesspools *below* them are safe. It should be understood that the words "above" and "below" refer in this connection, not to the slope of the surface, but to the direction of the underground flow. But even with this understanding the proposition is not by any means always true. The act of pumping a very large volume of water from a well, i.e. drawing more than its normal flow from it, without doubt alters the above conditions. The effect would be not only the lowering of the water in the well itself, but also the lowering of the surface of the underground water sheet for a certain distance all around the well. This distance, as is now well known, increases with the amount of pumping, and in this way sources of contamination, like outhouses or cesspools, which ordinarily might be beyond the influence of the well, may be brought within the zone of contamination.

The question whether a well may or may not become contaminated by a cesspool is, therefore, not merely a question of distance and of depth, but the means for drawing the water from the well must also be considered. With an old-fashioned bucket, drawing only a few gallons at a time, the danger may be very remote, whereas if a large pumping engine be put over the same well, the risk may become such as to render its use prohibitive.

The usual methods used by chemists in determining whether a well water supply has been contaminated need not be considered here. The fact whether a connection exists between a cesspool and a well may be established by several tests, one of the simplest being the salt test, in which a large quantity of common salt is thrown into a cesspool, the presence of which in the water of the well can readily be ascertained by a chlorine test. Another test consists in the use of lithium, and still another in the use of fluorescent " uranine," which gives to the well water a bright aniline green color. Some chemists have used "saprol" in order to detect the possible entrance of cesspool liquids into wells. It is claimed that such creosoliferous disinfectants may be detected by smell with a dilution of 1 in 1 million parts, and by taste when diluted to 1 in 2 million parts.

Care should be exercised in so arranging dug wells that they are well protected against surface pollution by providing at the top a tight and impervious covering to shed all surface water and by giving the upper end of the well a water-tight lining or wall to a depth of several feet, as shown in Fig. 3.

Driven or Tube Wells. — Driven or tube wells are made by driving iron pipes with pointed lower end, or with a shoe attached to the pipe, into a water-bearing stratum at a moderate depth from the surface. In the majority of cases they are much safer than dug wells, because there is not the same amount of danger of pollution by surface leakage. On the other hand, it sometimes happens that a driven well yields water unfit for drinking, because the water comes from an already polluted underground water stratum.

Where large volumes of water are required, and where the level of the water can be maintained at a practical suction distance, several wells are driven and connected together by horizontal piping, thus forming a battery of wells (see Fig. 6). The manner of connecting a series of wells by a suction main to a pumping station is shown in Fig. 7. Where a system of driven wells, all connected to one suction line, is used, there is apt to be trouble with air, and the greatest care is required to make the suction pipe line perfectly tight. It also happens at times that one well may rob the adjoining one of water, particularly if the rate of pumping is excessive.

Deep Wells. — Deep wells, sunk in sandy or gravelly soil, are driven or bored with tools similar to an auger, or else a casing is forced through the earth, into which the well pipe is inserted after the proper depth has been reached. In other cases, particularly in rocky strata, driven wells are drilled with special well tools or chisel drills, which are alternately raised and allowed to fall, being at the same time rotated.

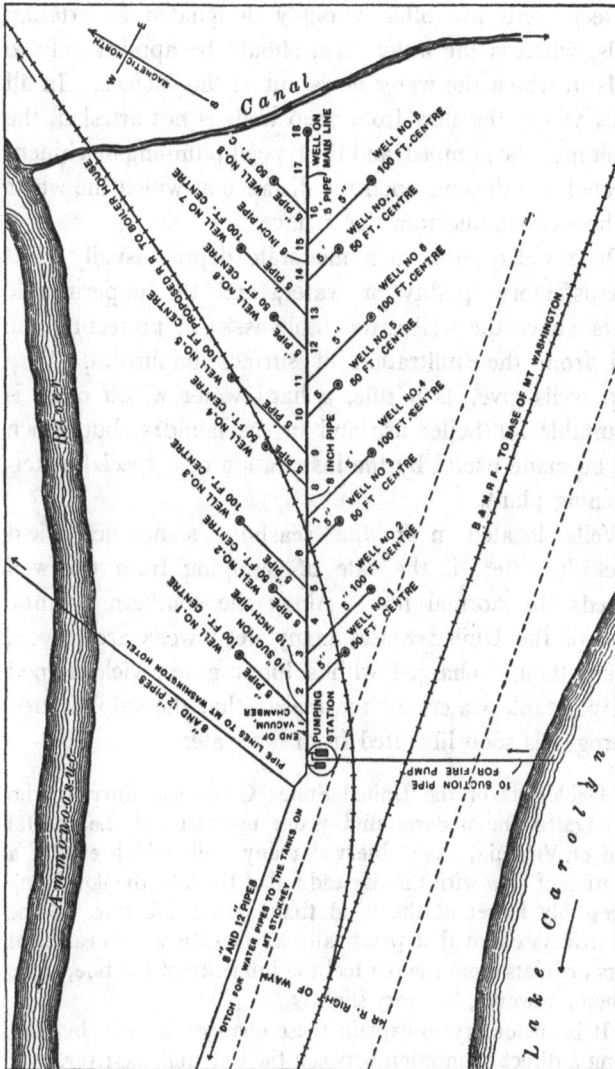

FIG. 6.—WATER SUPPLY OBTAINED FROM A SERIES OF WELLS.

Deep wells are often wrongly designated as artesian wells, whereas the latter term should be applied only to wells in which the water flows out at the surface. In all cases where the flow from deep wells is not artesian, the water must be pumped, and the type of pumping machinery selected will depend upon the distance at which the water in the well stands from the surface.

Deep wells, sunk to a moderate depth, usually yield a satisfactory quality of water, and the impermeable strata above the well water table assist in protecting the well from the infiltration of surface impurities. Very deep wells give, as a rule, a hard water which often is unsuitable for boiler use and in the laundry, but which can be made useful by the installation of a special water-softening plant.

Wells located near the seashore sometimes yield brackish water, if the rate of pumping from the well exceeds the normal flow. Along the southern Atlantic coast of the United States many deep wells are driven, which, though charged with sulphur gases, yield a good and palatable water, for the reason that the sulphuretted hydrogen is soon liberated from the water.

" Geologists of the United States Geological Survey, who investigated the underground water resources of the coastal plain of Virginia, have observed many wells which exhibit a variation of flow with the rise and fall of the tide, the flow being perceptibly larger at the flood than at the ebb tide. Some well drillers claim that practically all flowing wells near tidal rivers or inlets from the sea feel the influence of the tide, some of them, however, but very slightly.

" It is customary to explain these changes in yield by supposing a direct connection between the well and the river, lake or bay, but in many places, as in Eastern Virginia, such

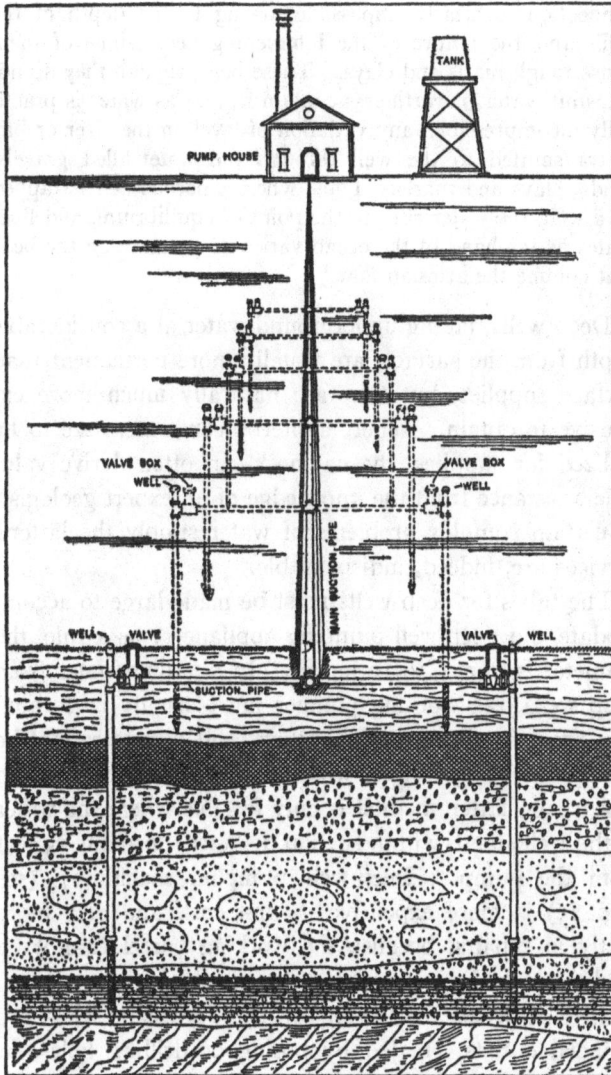

FIG. 7. — CONNECTION BETWEEN A BATTERY OF WELLS AND PUMPING
STATION.

connection is clearly impossible, owing to the depth of the wells and the nature of the intervening beds, some of them dense tough marls and clays. These beds, though they do not transmit water, nevertheless contain it, and as water is practically incompressible, any variation of level on the river or bay is transmitted to the well through the water-filled gravels, sands, clays and marls. Thus, when a porous bed is tapped by a well, the water rises to the point of equilibrium, and fluctuates as the hand of the ocean varies its pressure on the beds that confine the artesian flow."

Deep wells, taking underground water at a considerable depth from the surface, are usually more permanent than surface supplies, but they are naturally much more expensive to obtain. Where subterranean waters are to be utilized for supplies, the engineer can often derive valuable assistance from the knowledge of an expert geologist; in certain complex problems of water supply the latter's services are, indeed, indispensable.

The tubes for deep wells must be made large to accommodate the deep-well pumping appliances; as a rule, the diameter is from 6 to 10 inches. In pumping water from deep wells, the working plunger must usually be located at a great depth below the surface, and pumps cannot be attached to deep driven wells in the manner shown in Fig. 4; the latter method is only feasible where the water in the well stands within suction distance, or about 25 feet from the pump. Deep-well pumping machinery (see Fig. 25) is more expensive in first cost, and also more costly in repairs, because the working barrel cannot be readily reached.

Artesian Wells. — Flowing or true artesian wells are artificial outlets, created by boring or drilling holes, in which the water rises to and above the surface (see Fig. 8).

In a recent "Water Supply and Irrigation Paper" of the United States Geological Survey, Mr. Fuller discusses the meaning and significance of the word "arte-

FIG. 8. — A FLOWING OR TRUE ARTESIAN WELL.

sian," about the use of which there is considerable diversity of opinion, even among scientists and hydrogeologists.

The term "artesian" was originally used in connection with flowing wells, obtained in the province of Artois in France. In recent scientific literature, both in Europe and in this country, the term has been used promiscuously. But, as Mr. Fuller states, "the predominant scientific usage of the term is for all wells in which the water *rises* above the surface, in other words, for those wells exhibiting the hydrostatic or artesian principle. In popular practice it is applied, in addition to the use previously mentioned, to deep wells in general, especially those in rock, and to a certain extent to any drilled wells yielding water of good sanitary quality."

Mr. Fuller discusses in his paper the arguments for these various uses, and then gives the following definitions, agreed upon by the Division of Hydrology of the Geological Survey:

"*Artesian Principle.*

The artesian principle, which may be considered as identical with what is often known as the hydrostatic principle, is defined as the principle in virtue of which water, confined in the materials of the earth's crust, tends to rise to the level of the water surface at the highest point from which pressure is transmitted. Gas as an agent in causing the water to rise is expressly excluded from the definition.

"*Artesian Pressure.*

The pressure exhibited by water confined in the earth's crust at a level lower than its static head.

"*Artesian Water.*

That portion of the underground water which is under Artesian pressure and will rise if encountered by a well or other passage affording an outlet.

"*Artesian Basin.*

A basin of porous bedded rock in which, as a result of the synclinical structure, the water is confined under Artesian pressure.

"*Artesian Slope.*

A monoclinical slope of bedded rocks, in which water is confined beneath relatively impervious covers, owing to the obstruction to its downward passage by the pinching out of the porous beds, by their change from a pervious to an impervious character, by internal friction, or by dikes or other obstructions.

"*Artesian Area.*

An artesian area is an area underlain by water under artesian pressure.

"*Artesian Well.*

An artesian well is any well in which the water rises under artesian pressure when encountered."

"*Blowing Wells.* — A curious phenomenon is sometimes witnessed in the so-called 'blowing' or 'breathing' of wells. Hydrologists of the United States Geological Survey have in recent years observed that many wells emit currents of air with more or less force and sometimes accompanied by a whistling sound.

"Examples of this type of well were found in the State of Nebraska and in Southern Louisiana.

"It is explained that this blowing of air from wells is chiefly due to changes in atmospheric pressure or else to temperature changes. While the barometer stands low, air is expelled from the wells, and with a rising barometer the blowing becomes less and less until the current is finally reversed. Differences in the temperature of the surface air and the air in the soil produce similar effects."

"*Wells with Two Kinds of Water.* — Another curious phenomenon is observed in a flowing well at Logansport, Indiana, from which both fresh and sulphur waters are obtained. This well, located in Riverside Park, was drilled in 1905. An 8-inch pipe was sunk to a depth of 80 feet, and inside of it a 5-inch casing was placed. Fresh water from a limestone bed comes up between the two pipes, while water having a strong taste and odor of hydrogen sulphide comes up through the 5-inch pipe from a lower stratum in the limestone. The sulphur water flows with a volume of about one gallon per minute while the fresh water flows in a somewhat smaller quantity. The only other well of this kind known is said to be situated about 15 miles north of Cincinnati, Ohio, but this well is non-flowing."

"*Freezing of Wells.* — A further curious phenomenon, which causes much trouble in districts of the Northern States in

which air temperatures frequently go considerably below zero,
is the 'freezing' of wells, so much so that difficulty is often
experienced in keeping the wells open for use during the winter.
The shallow open wells give less trouble than the deeper,
drilled or double-tubed driven wells, in which the inner or
pump tube is carried below the outer casing. Wells in the
State of Maine, which are in granite, slate, or other compact
and close-grained rocks, do not exhibit the phenomenon of
deep freezing. But in Minnesota, North Dakota and Nebraska,
wells which penetrate porous deposits or cavernous limestones
having openings or passages through which the air can circulate, freeze every winter. The trouble occurs occasionally
in wells in Wisconsin, Michigan, Iowa, Missouri, Kentucky,
and Indiana.

"Deep wells which freeze sometimes exhibit the phenomenon
of indraft and outdraft of air, called the 'blowing' mentioned
heretofore, and also show fluctuations of water level, or in
flowing wells, changes in the discharge.

"The reason for the peculiar behavior of these wells is
assigned by geologists to barometric changes. Freezing,
indraft, low water level, small discharge and yield of clear water
occur during clear weather and a high barometer, whereas a
low barometer is accompanied by a thawing of the well, a
stronger yield and the occurrence of a discoloration of the water.
The direct cause of the freezing is an indraft of cold air when
the barometer stands high.

"To prevent the freezing of wells the following precautions
should be observed: —

"In open wells, where air obtains access through the soil
and at the junction of curb and cover, a cement cover should
be fitted tightly to the curb, and the curb itself should be
coated with cement for some distance below the surface.

"In drilled or double-tubed wells the current of cold air
drawn in at periods of high barometer between the outer and
inner casings near the surface and passing out in a porous bed
at the bottom above the water level will cause freezing if the
water is pumped so that it stands in the inner tube above the

lower end of the outer casing, and a long-continued current of such cold air may cause freezing of the ground water about and in the well tube. For this condition it is suggested that the space between the outer and inner tube near the surface be packed with some impervious material. A filling of cement resting on an improvised plug is probably the most effective. The home-made rag packing some times used is too porous to serve the purpose.

"The same treatment is suggested for wells with leaky casings, for driven wells passing through rocks porous enough to permit the passage of large currents of chilled air during periods of high barometer, and for wells in which the outer casing ends in some cavern or open passage; that is, the space between the well tube and the pump tube near the surface should be tightly plugged with impervious material. About some wells the ground crevices through which the air circulates are so numerous that immunity from freezing can be obtained only by plugging the space about the pump tube from top to bottom with cement."

Specifications for Wells. — It frequently happens that owners of country houses, having decided upon a well supply, obtain estimates for such a well from well drivers. The following specifications for driven, bored or drilled wells should be consulted before letting a well contract, and might with advantage be made a part of the same.

INSPECTION OF SITE. The contractor is to inspect the site for the proposed well, to note all the conditions of the problem, and to take into consideration all the necessary arrangements with regard to working space, headroom, transportation of his machinery, storage of same, and the provision of the necessary water, steam and fuel supplies.

EXECUTION OF THE WORK. The contractor to whom this work is awarded must begin operations ten days from date of awarding of contract.

He shall prosecute the work as rapidly as possible consistent with a first-class job, and without any unnecessary delay until the same is completed and accepted by the engineer.

He shall place the work in immediate charge of a competent and skilled foreman, who shall remain on the work until its completion except he be discharged for cause. The contractor shall also give the work his immediate personal attention, verify the records kept by the foreman, and note any unusual conditions which may be met with.

PREPARATIONS FOR THE WORK. The contractor shall make all his preparations for the well driving or boring immediately after the contract has been awarded to him. He shall ship his well machinery, all required tools, appliances and implements as soon as possible, so that he may begin work on time, and he shall also provide the required fuel for his boiler, also the water supply.

TOOLS, MACHINERY, IMPLEMENTS AND APPLIANCES. The contractor shall, at his expense, furnish and deliver to the site all tools, machinery, etc., required in the construction of the well boring, also all necessary planking, blocking, tackle, and the like.

He shall bear the expense of all transportation connected with the delivery.

FUEL AND WATER. It shall be understood that the contractor shall provide and furnish his own fuel, coal, wood or oil. Likewise shall he provide his own water for boiler or other purposes.

LOCATION OF WELL. The well is to be located where indicated on plan or at site selected by the engineer, and boring or drilling operations must be started directly over the selected spot.

DIAMETER OF WELL. The well shall be begun with a diameter of .... inches, inside measurement, and shall be continued this diameter to a depth of .... feet from the surface.

It shall be sunk, driven, drilled or bored truly straight, vertical and round.

Unless the bored or driven well passes through rock strata, it shall be understood that the contractor shall provide a well tubing, of .... inches inside diameter.

Where the drilling or boring is through rock strata, the directions of the engineer regarding providing or omitting tubing shall be followed by the contractor.

DEPTH OF WELL. The depth of the well shall be decided by the engineer after conference with the contractor. The contractor shall stop work at any point or depth designated or determined by the engineer.

DECISION AS TO METHOD OF CONSTRUCTING WELL. The method to be pursued by the contractor in constructing the well shall be determined by the engineer in charge, after conference with the contractor. In no case shall driving, drilling or boring appliances be used which, in the opinion of the engineer, are unsuitable for the work, or which are likely, if used, to diminish the sources of water, or which would interfere with, or prevent, the well tubing from being so fixed as to permanently exclude foreign matter or surface water, or undesirable courses of water at a lower depth.

WELL PIT. If the well is to be started in a well pit, it shall be constructed as per drawings, with brick or stone walls, of ....feet depth, and .... feet in diameter. The sides of the pit shall be finished inside with a coating of pure Portland cement, and the brick or stone walls shall rest on a concrete foundation. From the bottom of this pit the bore for the well shall be started in the manner directed.

WELL TUBING. The well tubing shall consist of standard galvanized wrought iron pipe. The diameter of the tubing shall be .... inches. All the joints of the tubing shall be screw joints tightly put together.

Where two concentric tubes are to be used, the internal diameter of the inner tube shall be not less than .... inches, and that of the outer tube not less than .... inches diameter.

The contractor shall determine the quantity of tubing required to comply with the requirements of this specification, and he shall provide the tube in ample quantities to enable the construction of the well to proceed continuously and without any hindrance. No allowance shall be made for any tubing except that actually inserted into the well.

If requested to do so, the contractor shall prove to the engineer's satisfaction that all foreign matter or water veins encountered during driving of the well have been efficiently and permanently excluded from the well.

RECORD OF PROGRESS. The contractor shall keep an accurate record of the weekly progress of the work. In it he shall describe the geologic strata through which he passes, the depth reached, and the amount of water furnished by the well, if any.

He shall furnish the engineer with a copy of the record, and also with samples of the various strata, which he shall label and name, and he shall also prepare a colored section showing each stratum, with its depth and the position of the well tubing.

FORM OF WELL RECORD. The following form of well record is suggested:

The well is located .... miles in a .... direction from the railroad station at .... in the town of ...., county of .... Owner's name .... Contractor's name .... Engineer .... Date when well driving was begun, ... Date when well driving was completed, ... What kind of rig was used? .... (cable, jet or ......). Diameter of well at surface or at top of well pit .... inches, reduced to .... at a depth of .... feet from surface. Total depth of well .... Length of tubing or casing inserted .... Main water supply struck at a depth of .... feet. Well pumps .... gallons (U. S.) per minute from depth of .... feet. Well flows .... gallons (U. S.) per minute. If flowing well, what is the hydrostatic pressure? .... Recorded by .... Address ....

TESTING THE WELL. Whenever directed by the engineer, the contractor shall arrange for making a pumping test of the well to ascertain the yield of water of the well at any depth.

The contractor shall state in his estimate the cost for each trial test. He shall furnish the required pipes and fittings, and the test pump, and make all connections required for such test.

The test pump shall be not less than 4 inches in diameter inside, with a 24-inch stroke, and pump rods, fittings, couplings, etc., shall be of ample strength for raising the water from any required depth to a point not less than 4 feet above the level of ground at well. He shall also provide at least two 50-gallon barrels for measuring purposes, also the necessary discharge pipe and hose to carry the water from the pump to the barrels. He shall arrange for the alternate filling of the barrels.

If so required by the terms of the contract agreement the contractor shall make at least two tests of the well at his own cost.

The contractor shall provide for the tests a steam or other engine of not less than 8 horsepower, together with all necessary accessories. He shall provide the required steam or fuel for running the engine.

If the test should indicate that the desired amount of water cannot be obtained from the depth of well when test is applied, the contractor shall continue the boring when directed to do so by the engineer. Any additional work done before the receipt of such instructions from the engineer is at the contractor's own risk.

REPETITION OF TEST. After the well has reached a certain depth, to be stated in the contract agreement, the contractor shall make tests of the well at each additional 25 feet in depth of well.

DURATION OF TESTS. Each test of the well shall be continued by the contractor for a period of not less than 10 hours unless contrary instructions are issued by the engineer.

All tests shall be made in the presence of the engineer or his representative.

The final test of the well shall be continuous for a period of 24 hours (night and day).

AMOUNT OF WATER EXPECTED. The supply of water required (in any case not less than .... U. S. gallons per hour) is the greatest quantity the well is capable of yielding when pumped from a depth of not less than .... feet with a stroke of .... inches, running .... strokes per minute, and delivered into barrels placed .... feet above the surface of the well bore and .... feet horizontal distance from it.

FAILURE OF TEST. The indicated horsepower or brake horsepower of the engine driving the pump, and the cost of the consumption of steam or fuel per 1000 U. S. gallons raised from the well, when pumping the maximum quantity of water, shall be clearly stated by the contractor.

In the event of the pumping plant proving incapable of producing the results required, or upon the failure to maintain the guaranteed efficiency and cost of working, the engineer shall have the right to order the removal of the pump without compensation to the contractor.

LOSS OF WELL-DRIVING TOOLS. Should any well-boring tools or any part of the machinery be broken, lost or injured during the driving of the well, necessitating the abandoning of the well, the contractor shall begin anew another boring, and he shall not be entitled to any compensation for the driving of the abandoned well. The contractor shall continue a new well at his own expense until he can turn over to the owner a well conforming to the requirements of the specifications and the contract.

TORPEDOING THE WELL. The contractor shall not "torpedo" the well boring without consultation with the engineer, nor without his written consent to do so.

REMOVAL OF DEBRIS. Upon the completion of his contract, the contractor shall remove all débris and material resulting from the well driving operations.

REMOVAL OF WELL MACHINERY. When directed by the engineer to do so, the contractor shall remove from the owner's premises all well-driving machinery and well tools, and he shall effect all removal at his own expense.

COMPLETED WELL. On completion, the top of the well shall be provided with coupling or socket and with plug, and both shall be fastened together with Yale padlock in such a way that the well cannot be tampered with.

The keys for the padlock shall be turned over to the engineer before final certificate of payment can be issued.

TIME FOR COMPLETION. The contractor shall be bound under his contract to complete the well to a depth of ..... feet within the period of .... working days. Should a greater depth be required, an extension of time, amounting to .... days for .... feet depth of well shall be granted.

PAYMENTS. The contractor shall be entitled to a payment of .... per cent of the cost of the work so far done after the completion of each test. The payment shall be certified by the engineer.

The balance of cost shall be due to the contractor within .... days after completion and acceptance of the well.

DAMAGES. The contractor shall exert the usual and ordinary care not to cause damage to the owner's premises during his operations. For any damage done he shall be held responsible.

The contractor shall assume all risk of fire, explosion or other accident or damage to his machinery.

The contractor shall be held personally responsible for any accidents to persons, or damage to property, arising during and from his operations.

The contractor shall replace at his own expense any tools which may have become lost.

ESTIMATES. Bids for this work must be based upon a personal inspection of the locality, and shall state the price per foot in depth of the well. Where the well is of varying diameter, the price for each pipe diameter shall be clearly stated.

Where the well is to start from a well pit, the depth of well is to be understood as measured from the bottom of the pit, and not from the surface.

Water Finders. — Before digging or boring a well for water supply, it is often advisable to consult an expert hydro-geologist, or to refer to the numerous published reports of the United States Geological Survey.

Formerly it used to be customary to employ "water finders" or so-called "water witches" to have them designate the best spot for obtaining water. From an article on "Water Supply for Country Houses," contributed to a leading magazine by the writer, I quote the following:

"Water finders, being usually shrewd observers, locate by the aid of a hazel twig the exact spot where water may be found. The superstitious faith in the power of the forked twig or branch from the hazelnut bush to indicate by its twisting or turning the presence of underground water was at one time wide spread, but only the very slightest foundation of fact exists for the belief in such supernatural powers.

"In Europe, attention has again, during the past years, been called to this 'method' of finding water, and it has even received the indorsement of a very high German authority in hydraulic engineering, a man well up in years, with a very wide practical experience, and an author of the most up-to-date handbook on water supply, but men of science have not failed to contradict his statements." (See footnote, page 116.)

In this connection the following article on "The Use of the Divining Rod," taken from a bulletin of the United States Geological Survey, is of interest.

"Numerous devices are used throughout this country for detecting the presence of underground water — devices ranging in complexity from the forked branch of witch hazel, peach, or other wood, to more or less elaborate mechanical or electrical contrivances. Many of the operators of these devices, especially those that use the home-cut forked branch, are perfectly honest in the belief that the working of the rod is influenced by agencies — usually regarded as electric currents following underground streams of water — that are entirely independent of their own bodies, and many uneducated people have implicit faith in their ability to locate underground water.

"In experiments with a rod of this type, one of the geologists of the United States Geological Survey found that at points it turned downward independently of his will, but more complete tests showed that the downturning resulted from slight and — until watched for — unconscious changes in the inclination of his body, the effects of which were communicated through the arms and wrists to the rod. No movement of the rod from causes outside the body could be detected, and it soon became obvious that the view held by other men of science is correct — that the operation of the "divining rod" is generally due to unconscious movements of the body or of the muscles of the hand. The experiments made show that these movements happen most frequently at places where the operator's experience has led him to believe that water may be found. The uselessness of the divining rod is indicated by the facts that the rod may be worked at will by the operator, that he fails to detect strong currents of water running in tunnels or other channels that afford no surface indications of water, and that his locations in limestone regions where water flows in well-defined channels are rarely more successful than those dependent on mere guesses. In fact, its operators are successful only in those regions in which ground water occurs in a definite sheet in porous material or in more or less clayey deposits, such as the pebbly clay or till in which, although a few failures occur, wells would get water anywhere.

"Ground water occurs under certain definite conditions, and as in humid regions a stream may be predicted wherever a valley is known, so one familiar with rocks and ground-water conditions may predict places where ground water can be found. No appliance, either mechanical or electrical, has yet been successfully used for detecting water in places where plain common sense or mere guessing would not have shown its presence just as well. The only advantage of employing a "water witch," as the operator of the divining rod is sometimes called, is that skilled services are obtained, most men so employed being keener and better observers of the occurrence and movements of ground water than the average person." [*]

Collecting Galleries. — Another method of utilizing the subterranean water is to have a horizontal collecting gallery, built either of glazed earthen pipes of large size,

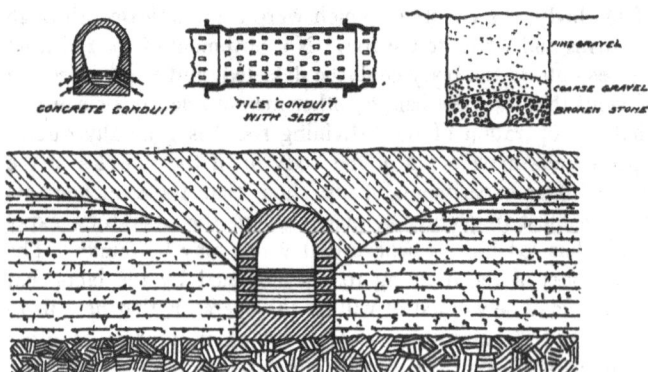

FIG. 9. — COLLECTING GALLERIES OR CONDUITS FOR UNDERGROUND WATER.

and provided on their circumference with numerous slots, or consisting of a brick conduit with numerous openings at the sides (Fig. 9), the bottom of the gallery being located

[*] See S. T. Child, Water Finding and the Divining Rod, Ipswich, 1902.

See also the following recent German pamphlets: —

Georg Franzius, die Wünschelrute, Zentralblatt der Bauverwaltung, Sept. 13, 1905. Georg Franzius, Meine Beobachtungen mit der Wünschelrute, Berlin, 1907. Dr. L. Weber, Die Wünschelrute, Kiel, 1905.

Fr. Koenig, Ernstes und Heiteres aus dem Zauberreich der Wünschelrute, Leipzig, 1907.

at the level of the impermeable stratum. Such a conduit for underground water should terminate in a brick well, from which the water is pumped, while any sand or gravel carried in the water is deposited at the bottom of the chamber.

Recent experience has shown that it is necessary, in the storage of underground waters, to keep away the light, as otherwise there is apt to be trouble from vegetable growths, which at certain seasons impart to the water a bad odor, a disagreeable taste, or both.

Rain Water and Cisterns. — In the country, rain water, as it falls from the clouds, is apt to be much purer than near large towns, although it always collects some impurities in its downward path. It may be collected for use, provided the water delivered during the first part of a storm, which is apt to be quite impure, is intercepted and run to waste. Proper precautions should therefore be taken to let the first washings from the roofs run away. If this is done, rain water may be considered, in the absence of springs or where a shallow well supply is polluted, a suitable but limited supply in the case of smaller houses or farm buildings. So-called self-acting rain-water separators may be arranged on the outside conductor pipes of the house, and in this way a pure supply can be secured, even with large buildings or institutions. In the case of the latter it is often a good practice to make use of the soft rain water for laundry and for boiler feed purposes.

Rain water is stored either in rain water tanks placed in the attic of houses, or else in underground brick or stone reservoirs, called cisterns. The underground tanks are preferable because they keep the water cooler and prevent vegetable growth, since the light is excluded. The cisterns should be built water tight to prevent leakage and contamination from a polluted soil. In determining the sizes of cisterns, the annual mean rainfall of the locality should be looked up from the meteorological records. After making due allowance for evaporation, the water supply available for storage may be calculated.

Instead of merely utilizing the limited roof areas, portions of the surface of the ground may be used if carefully prepared as "catchment areas" with concrete and cement. In the Bermuda Islands, for instance, the water supply of the towns and parishes is obtained from the rainfall exclusively, the porosity of the coral rock, which is the geologic formation of the islands, causing the rain to percolate so quickly as to render the use of wells impossible. Visitors to these islands have been struck with the fact that no matter how heavily it rains, the roads dry up entirely in a few hours, owing to the peculiar geologic formation. Neither subterranean waters nor springs being available, the houses are provided with an underground cemented storage tank, which is connected either with roofs finished with thin slabs or slates of the coral rock, or else the tanks are connected with so-called surface catchment areas, which are specially prepared water-gathering surfaces to which a good slope is given. These are sometimes quite extensive and are always enclosed with a railing to keep off the cattle. Underneath the artificially prepared area, and at the lowest point of the slope, an underground tank for the storage of water is built. The water supply of the entire islands is obtained in this way, and great care is exercised to keep the roofs and the catchment areas pure, clean and whitewashed.

Brooks and Streams. — Another source of water supply may be found in brooks or running streams. The taking of water from open streams is limited by legal restrictions,

FIG. 10. — A BROOK FLOWING THROUGH MANURED FIELDS.

much more so than is the case with subterranean waters. A knowledge of the "law of water" will prove useful to the engineer undertaking works of water supply in the

country, and it must be carefully considered before any
works are planned or undertaken.   These legal restrictions
will be briefly taken up further on.

In considering open water courses, whether brooks or

Copyrighted by Detroit Photographic Co.

FIG. 11. — UPLAND BROOK AS A SOURCE OF WATER SUPPLY.

rivers, as sources of supply, it should be remembered that
even in the country such streams are, in many cases, in
danger of pollution by the surface washings from manured
fields (Fig. 10).   In the case of the smaller streams, the

water is often found to be quite impure and unfit where settlements of houses, hamlets or villages are located above the proposed intake. This tendency to contamination increases with the growth of population along the banks. The water should therefore be improved by providing a filtering gallery between the stream and the suction well or reservoir, or else it should be passed through artificial sand filters. River water, unless properly filtered, is also objectionable as a source of water supply, because during floods it is apt to be not only quite turbid but also polluted with organic matters drained into the river from the surrounding cultivated farms.

Streams flowing through uninhabited and uncultivated upland regions and those from mountainous districts (Fig. 11) usually furnish a pure supply, but the smaller brooks are very liable to have a very changeable rate of flow, being mountain torrents immediately after heavy rainstorms and drying up almost or entirely in hot summer weather. For such reasons they are to be avoided as a permanent source of supply except where it is contemplated to store up a part of the flow in an artificial reservoir.

Lakes. — The water from many lakes is clear, bright, and potable, and may be used as a supply, provided there are no settlements along the shores which may cause pollution of the water. In mountainous, uninhabited regions lakes furnish very pure water and are, moreover, usually at an elevation permitting a gravity supply. As a rule, the larger and deeper the lake, the greater the likelihood of obtaining a pure supply, for all lakes act as settling basins for the sediment brought into them by the streams tributary to the lake, and their water then becomes purified by "quiescence."

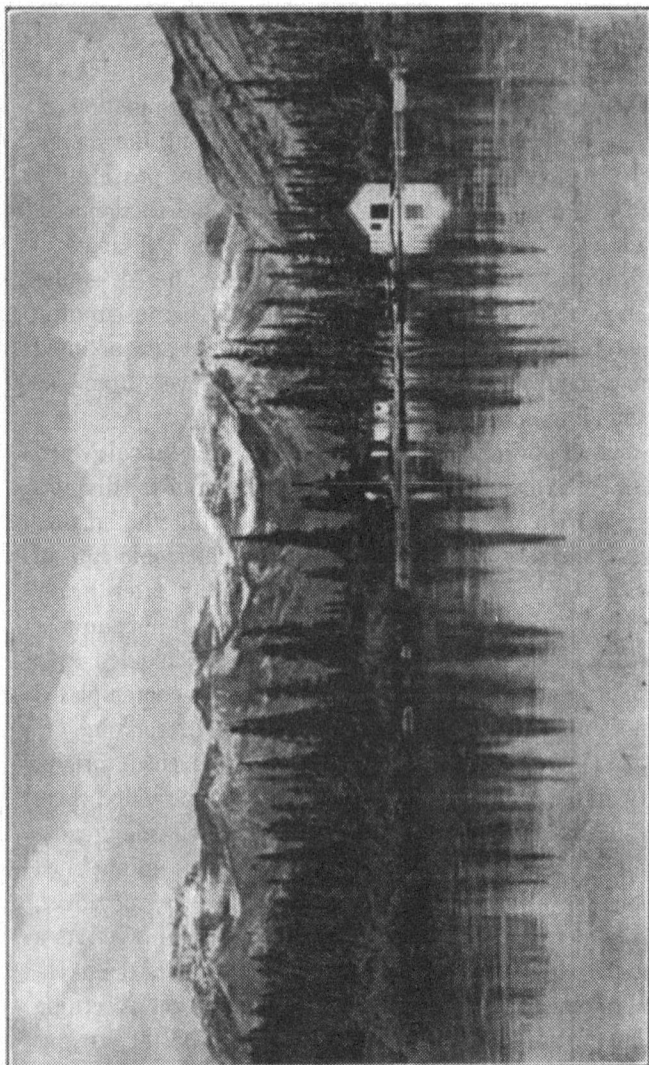

Copyrighted by Detroit Photographic Co.

FIG. 12. — MOUNTAIN LAKE AS A SOURCE OF WATER SUPPLY.

The smaller lakes, as for instance those located in the mountains in sparsely inhabited regions (Fig. 12), generally form an exceptionally pure supply and may be kept fit for use, provided a stringent sanitary supervision is exercised.

The water of lakes on the shores of which summer hotels, cottages, or camps are located, should be protected from pollution by the enactment of State laws, forbidding absolutely the common practice of depositing on the shores or in the water domestic sewage, garbage, liquid and solid excreta and wastes of any kind. This matter will be referred to again in the chapter on Sewage Disposal.

Impounded Surface Water. — Water from surface streams may be utilized by collecting it in an artificial storage reservoir, built in the watershed. Such impound-ing reservoirs, which usually permit the use of a gravity supply, are to be found in some mountain regions, where the land may be acquired cheaply and where a short dam may be constructed across a narrow part of the valley of the stream at its lower end. It is, of course, necessary that there be no settlements or manured fields on the drainage area forming the watershed of the stream.

If there are habitations near the watershed, its sanitary condition requires close watching, and even the temporary summer camps, or laborers' temporary huts or shacks, or the use of a part of the area for picnic purposes, should not be tolerated. In making such an artificial storage reservoir for surface water, the brushwood, vegetable matter, and the peaty soil sometimes forming its bottom, should be carefully removed.

In the case of a water supply from a "dammed-up" watershed, strict regulations may be necessary to guard against its pollution. This method of obtaining a supply is not often employed in the case of country houses.

Legal Considerations. — Enough has been said in the foregoing to show that the problem of finding a safe and sufficient source of water supply is often beset with difficulties from the *engineering point of view*.

In many water supply problems, and in particular in the selection of an available source of supply, engineering questions are not the only ones to be dealt with, for *legal considerations*, which must be thoroughly understood and respected, may govern or affect the problem.

The laws on the subject seem to vary in different countries. There appears to be a sharp legal distinction between waters flowing over the surface without a clearly marked channel, those which flow in an open natural water course, and underground waters. Briefly stated, the water flowing in an open stream is not the exclusive property of anyone. Its use, subject to certain restrictions, is given by law not only to the owner over or through whose land the water flows, but to every one having a right of access to it. In other words, all owners of land abutting on a stream enjoy certain privileges of this stream, which are known as the "riparian rights." Thus, every one of the riparian owners of a stream has a right to the *ordinary* use of the water flowing along his premises. He may use the water from the stream for the domestic supply of his house and also for his cattle and horses, and in this way he may diminish its volume of flow *to some extent*.

He has the right to *extraordinary* uses of the water, such as for water power, irrigation or manufacturing purposes, only provided that by such use he does not change its character, or diminish its flow, or make it less useful to the other owners, and in this way encroach upon the legal rights of his neighbors above and below stream. He cannot, for instance, intercept the regular flow of water in the stream for the purpose of a water supply of a settlement of summer cottages to such an extent as to interfere thereby with the rightful uses of the same water by the other riparian owners.

Again, he must not discharge domestic or manufacturing sewage into the stream and thus render the water impure or polluted. Those living further downstream must not, by his acts, be robbed of the ordinary use of the water, which they as well as he require for water supply in the home, for the farm, etc. The rights conceded to them by law can be interfered with only where there is a mutual understanding, or where a legally sufficient compensation is given.

On the other hand, in the case of underground water, no such rights of neighbors are entertained by law, hence anyone may dig or drive on his own property a well of any depth, to tap the underground sheet of water, irrespective of the fact that by doing so he lowers the water in the well or in the spring of his neighbors. An exception to this rule is made by law only in the case of subsurface water which flows "in a defined channel."

Unless surface waters constitute by law a stream with well-defined channel confined between two banks, they are not subject to any legal restrictions.*

* See Appleton's Universal Cyclopaedia, Articles on Riparian Rights, Water Courses, Filum Aquae, Lakes, Seashore, Law of Rivers. See also E. B. Goodell, Review of Laws Forbidding Pollution of Inland Waters in the United States, published by the United States Geological Survey, 1904, Water Supply and Irrigation Paper No. 103.

See also D. W. Johnson, Relation of the Law to Underground Waters, published by the U. S. Geol. Survey, 1905, Water Supply Paper No. 122.

From the foregoing it will be seen how necessary it is, in the case of country water supplies, for the engineer and the owner to become thoroughly acquainted with the legal aspects of the case. In the plans for water supply for country settlements, it will often be found advisable or necessary to arrange suitable terms with the owners of the land on which a spring is situated, and in other cases with the riparian owners of a stream, from which it is proposed to take a water supply.

Rights of Way and Easements. — In the case of gravity supplies, it sometimes becomes necessary to run conduit pipes across land belonging to other owners, and this cannot be done unless an easement is first obtained, which gives to one person the privileges of right of way and also the right of entry for necessary repairs. The person who obtains the easement acquires thereby also certain rights against the neighbor, for instance the right that he can enjoin him from building houses or planting trees over the line of easement. It would, therefore, seem necessary that a proper compensation should be given for the rights acquired by the easement.

Gravity and Pumping Supply. — The sources of supply considered in the foregoing may be located either above or below the building to be supplied. If located at an elevation above the building, suitable and sufficient to give a good flow without having recourse to pumping, a gravity supply can be arranged for if surface water, springs, lakes, or brooks form the source of supply. Inasmuch as the gravity system requires no running expenses once the supply conduit has been constructed, it is in many cases given the preference over a pumping system. A supply from a well or from a rain-water cistern generally implies pumping of the water. The distance between the source of supply and the building has, of course, an important bearing upon the question of first cost. As a rule it is best not to exceed a certain distance in length of gravity conduit, but rather to select a nearer source and install a pumping plant.

Quality of Water. — Where only a single source of supply is available, it is necessary to carefully determine the quality of the water, because this may require special means for its purification, and in case it is a very hard water, special methods and appliances for softening the water. All water containing even a small amount of organic matter of animal origin must be considered "suspicious," and a large amount of organic substances present should lead to its being condemned as "dangerous."

Water Analyses. — When several sources of supply are available, the selection is sometimes made after obtaining a comparison of the physical and sanitary qualities of the water sources. It is necessary, in all investigations of this character, to examine not only the physical characteristics of the water, such as taste, color, odor, temperature, turbidity and hardness, but also to subject a sample, or preferably several samples, of the water both to chemical and bacteriological analyses, in order to test its suitability from a sanitary point of view. Modern knowledge on the subject of water supply requires that the results of the different analyses should be taken together, and much more stress than formerly is laid upon the number and kind of living organisms, or bacteria, found in the drinking water.

Sanitary Inspection of Source of Supply. — In addition to the tests mentioned, a sanitary inspection of the source of supply, of the watershed, of the river basin and its tributaries, or of the lake and its water sources, should be made. In the case of running streams it is considered insufficient merely to take a sample of water at the place where it is intended to locate the suction inlet to the pumps. It is very much better to take additional samples above such point, and incidentally to determine, by an inspection, the character of the country through which the stream runs. It will be found, in many cases, that the water course is made use of, by settlements located upstream, as a convenient outfall into which their sewers or drains empty. One must, therefore, always ascertain not only the present purity of the supply, but likewise its safety from future pollution.

Quantity. — A source of supply may be ever so favorable as regards its quality, yet if the quantity available falls short of the volume required, it may become necessary to discard it and to look for other sources. In the case of small springs, the quantity yielded can usually be readily measured, and small brooks or water courses can be gauged by means of weirs. For large and important buildings, it is necessary, when a supply from wells is desired, to drive test wells and to ascertain the supply by means of pumping. In this way the quantity available is readily determined, and the next step is to compare this with the quantity actually required.

Amount of Water Used. — Right here it is well for the engineer to remember that the allowance in the case of country mansions must be generous and ample, for in the buildings and on the grounds of large country estates water is apt to be used very lavishly. A fair allowance, so far as the house itself is concerned, for water used in drinking, cooking, washing, bathing, ablutions and flushing closets would be 50 United States gallons per head per day. But the number of persons who may occupy the building, or the prospective population of an institution, often have to be guessed at in making an estimate; this at once introduces a factor of uncertainty. While in cities the average daily consumption reaches nearly 100 gallons or even more, of which a large quantity is really wasted, I find that American public institutions, forming a group of detached buildings in the country, sometimes use enormous quantities, such as 200 gallons and over per capita. It is true that a part of this consumption is intentional waste, caused by leaving faucets open to prevent the freezing of poorly located plumbing, and by a too abundant use of water in the flushing of fixtures or in bathing. The other part is doubtless due to carelessness in not keeping the water fittings of houses, the faucets, ball cocks and cistern valves, in proper repair.

In the case of large country mansions or estates, we must add to the inside or domestic consumption the large volumes of water required for stable use, for the watering of horses and cattle, for the washing of carriages and harness, for the cow barns, the dairy, etc. The allowance for these items alone sometimes amounts to from 50 to 200 gallons per day for each horse and carriage. A further large amount must be provided for the watering of the roads, for the sprinkling of the lawns and flower beds, for orchard and vegetable gardens, and for the use in the conservatory, which amount rarely falls short of 1000 and even 2000 gallons per day. Sometimes an average of 500 gallons per acre of ground is allowed. Where ornamental fountains are provided in gardens, these also consume a large supply, varying, according to size and number of fountain jets, from 50 gallons to several thousand gallons per hour.

Finally, it is necessary to provide, in the storage of water, a large volume to be drawn upon in the case of an outbreak of fire. The danger of a conflagration is, of course, ever present, and it becomes serious in proportion to the remoteness of the buildings from the nearest village or town fire department.

Water Pressure. — The volume of supply for which provision must be made should be available at the buildings and on the grounds under a suitable, i.e. sufficient pressure. This brings up the engineering question: what is a suitable pressure of water to be provided? In order to answer this question intelligently, it is necessary to discriminate between domestic-service and fire-service pressure. Domestic pressure means the pressure required to cause the water to flow at the highest faucet in a house under a moderate but still sufficient flow; or, where there is an attic tank, the pressure which will constantly keep the tank supplied. Fire-service pressure, however, means a much higher pressure, for in order to be satisfactory and effective, fire streams should be able to reach above the roof of a house with some force.

In view of the usual height of buildings in the country, it is considered by engineers that the pressure of water sufficient for fire-extinguishing purposes should be from 40 to 50 pounds per square inch at a hydrant in the basement. In the case of large hotel buildings, the minimum requirements of fire underwriters are somewhat higher, namely 60 pounds in the basement or the ground floor. In some cases, such a pressure is furnished by gravity supplies (see example No. II); in other instances, even where a gravity supply exists, the water must be pumped to give a sufficient head or pressure. The latter is usually attained by providing large tanks, elevated on towers built of sufficient height. It should be remembered also that, where small fire hose is used, the friction in long lines of hose is very large, absorbing much of the available pressure, and thereby reducing the height of the fire stream to a considerable extent.

The cases are exceptional where a natural or gravity pressure is *too heavy* and where in consequence, in order to avoid the excessive straining of the house pipes and fittings, the pressure has to be reduced. This is accomplished by the use of pressure regulators or pressure-reducing valves. It may be said, however, that the use of these is not to be recommended, except where they are absolutely necessary, for even the best of these valves cause occasional trouble by getting out of order.

Supply System. — Having determined the source, the quality and the quantity of water, and the desired pressure at the building in the manner outlined above, the engineer is enabled to decide upon and lay out a supply system, which in the special case under consideration should be looked upon with greatest favor, as being the best, most sanitary and most economical, both in first cost and in cost of maintenance. In general, it may be stated that a gravity supply always seems to be the most desirable; it is sometimes, though not always, the cheapest in regard to maintenance and running expenses. A pumping system requires either manual labor or skilled attendance for the pumping machinery and is, besides, liable to give trouble on account of the breaking down or the wear and tear of the latter.

A gravity system may be installed with or without storage tanks, but it is considered preferable to provide the latter. A pumping system requires the use of reservoirs, located on suitably high elevations. Where these are not available, elevated tanks or standpipes are used to secure the required storage and pressure. Sometimes underground pressure tanks are substituted for reservoirs or elevated tanks, and in a few cases pumps are arranged to discharge directly into the supply mains to the buildings, and are then provided with relief valves, opening up when the pressure becomes excessive. The latter modification does not provide any storage for domestic and fire purposes, and hence requires the pumping plant to be laid out in duplicate.

Pumping Water. — In all cases where water cannot be supplied by gravity, artificial means for raising it from a lower to the desired higher level must be provided. This leads to a more detailed consideration of the usual types of water-pumping machinery.

All such machinery consists essentially of two parts, namely the pump lifting the water, and the motor operating the pump. There are numerous varieties of apparatus for raising water from which to choose, and only the chief kinds will be mentioned here. In actual practice, the peculiar conditions governing each problem usually restrict the choice of suitable pumping machinery to a few types.

Motors. — As regards the motive force or power, we distinguish between motors operated by *natural* and those using *artificially created* forces. The natural forces applied for pumping water comprise the muscular or hand power of men, the power of animals, such as horses, oxen, or donkeys, the force of the wind, the force of water in motion, and the power of falling water. Artificial motive powers embrace all motors using fuel in some form of engine, in which heat is converted into power. The fuel may be either coal, wood, gas, oil, naphtha or gasoline, and the power derived therefrom may be either steam, compressed air, hot air, or electric power.

Simple Water-lifting Devices. — Water may be raised by other and simpler devices than pumps, such as machines with scoops, pails or boxes, machines with moving water channels, or finally those in which water is raised in fixed conduits by chains or other means. Many of these devices are of a primitive kind, have gone almost out of use, and are only historically interesting. We may safely pass them over to discuss the more universally used machines, in which water is raised in fixed channels by means of pistons, plungers, impellers, worm gears, or by steam, water, or compressed air jets. Such machines are designated as *pumps*.

Pumps. — Pumps may be either lift pumps or lift and force pumps. We can also distinguish, according to the details of the pumping machinery, piston and plunger pumps, single and double-acting, rotary and reciprocating, and finally simplex, duplex and triplex pumps. In the last-named pumps, the three pistons or plungers are mounted on the same shaft, while the cranks are set at equal angles to each other; a much steadier and more uniform discharge is thereby secured. We may also distinguish between vertical and horizontal, stationary and portable forms of pumps. Water-lifting apparatus also differs, in so far as it may be adapted either for surface supplies and shallow wells or for deep wells.

As regards the connection between the motor and the pump, the power in many machines is applied directly, as in the direct-acting steam pump, and in the direct-connected electric pump, both motor and pump being mounted on the same shaft. In other cases, the power is applied indirectly by means of gearing, belting, noiseless drive chains, or crank-shaft movements, and the pumps are then designated as power pumps.

To each kind of pumping engine belong certain advantages and disadvantages. In every case the conditions under which the pumping engine is to do the work should be carefully looked into before a selection is made. One should always consider not only the first cost of the pump but also its running expenses, the cost of the fuel, and of repairs and maintenance of the machinery.

Hand Pumping. — Hand pumping is uncertain and unsatisfactory; it can be applied only where small volumes of water are to be lifted, and where the lifts are small. Lift or force pumps operated by hand are useful only in the case of farm buildings or cottages of moderate size. Their only advantage consists in the cheapness of the labor. In countries where animals are cheap, these too are used to some extent in operating water-lifting apparatus, but as a rule such use of farm horses or oxen does not commend itself.

The water to be used on farms or in farm buildings is more often lifted by the force of the wind, with windmills, or by the force of water in motion, with hydraulic rams. As will be pointed out below, both are very economical, as the motive power costs practically nothing. If neither of these is available, some form of fuel engine must necessarily be used. In this respect both the gasoline and the oil pumping engines are very popular, because the cost of the fuel which they require is very small.

Windmills. — Wind as a motive power is utilized extensively for the pumping of water, and in numerous cases windmills, when properly installed, provide a satisfactory water supply at a minimum expense (Figs. 13 and 14). This motive

FIG. 13. — WINDMILL FOR PUMP-
ING WATER TO TANK ON SAME
TOWER.

FIG. 14. — WINDMILL FOR PUMP-
ING WATER; TANK LOCATED
ELSEWHERE.

power is convenient as well as economical, but it is also somewhat uncertain, for the source of power is beyond human control, and the wind does not always blow with sufficient force to move a windmill. A velocity of the wind ranging from 4 to 8 miles an hour is not sufficient for larger wheels; the best results are obtained with wind velocities of from 12 to 16 miles an hour. Windmills require some skill and judgment in their application and erection, and a preliminary careful study of the topography of the grounds should always be made with a view of locating the windmill in the most favorable position, for it is essential to erect them in a rather exposed elevation, on high ground, so that the wheel may receive the full force of the wind. Even when so situated, a windmill will work only a part of the time, an average of 8 hours per day, and to allow for the unavoidable periods of calm, it is necessary to provide a very ample storage capacity in the tanks.

A complete windmill pumping plant comprises a tower for the wheel, a pump with piping, and a storage tank. The tank may be placed below the windmill on the same tower, or it may be supported elsewhere. The tower must be so placed as to secure the best exposure of the wheel, and towers which are too low frequently lead to failure. The wheel must be carefully mounted, and the tower must be substantial and strong, and be supported on brick or stone piers. The corner posts should be anchored to the piers by means of strong wrought-iron bolts; the tower should be framed and rodded, and not, like the cheaper ones, merely spiked together. Both wooden and steel towers are used, and there are wooden as well as steel wheels.

The wheels should be built with a view to withstanding hard storms. The pumping apparatus should be of the simplest kind in order to be easily managed by the farm hands. Tank, pump, and piping must be frostproofed, except in cases where the supply is designed for summer

FIG. 15. — IMPROVED FORM OF HYDRAULIC RAM.

use only. The wheel is usually provided with self-regulating attachments. Sometimes a windmill is used for other purposes in addition to the pumping of water, such as the running of dynamos for generating electricity.

Hydraulic Rams. — The force of running water is utilized in hydraulic rams for the lifting of water to a height greater than that of the source of the supply for the ram. Wherever there is a volume of flowing water ample at all times to run a ram, this is one of the most useful, cheap, economical and satisfactory motors for lifting water. A ram requires but very little attendance and hardly any care or expense in maintenance. It operates automatically and occupies a field entirely by itself.

The ram is set up in an underground pit located at a suitable distance below the source of supply; the fall required to operate a ram may be as little as 18 inches. The existing conditions as to fall and length of drive pipe must be first carefully studied out, for they govern the location of the ram, its size, and the dimensions of the drive and the discharge pipes. Hydraulic rams are made in many sizes and capacities, the smallest supplying about 500 gallons, and the largest up to 500,000 gallons in 24 hours. They will elevate water to a height from five to several hundred feet. Sometimes several rams are set to work in batteries, with separate drive pipes, but with a common discharge pipe (see Fig. 34). The improved forms of rams, or hydraulic engines, shown in Fig. 15 and also in Fig. 34 are much more efficient than the old forms and develop an efficiency of from 70 to 80 per cent. These types have adjustable valve mechanisms which can be readily regulated so as to secure the best working conditions. Double-acting rams are also manufactured, which use an impure source of water to pump a pure supply.

Water Wheels. — If running water exists in sufficient quantity, water may be pumped by utilizing the power of falling water either in water wheels or in turbines (see Figs. 16 and 17). Hydraulic power is economical when we consider the running expenses only, for these may be

FIG. 16 — DOWNSTREAM VIEW OF WATERWHEEL USED FOR PUMPING WATER

FIG. 17. — UPSTREAM VIEW OF WATERWHEEL.

almost nothing; hydraulic power is easily managed as the machines require only occasional looking after. On the other hand, it should not be overlooked that the initial outlay for hydraulic power, for dams, sluiceways, etc., may in some cases be quite large. Another drawback is that in many situations water power is unreliable and at times insufficient.

Water wheels always turn on a horizontal axis, whereas turbines are usually mounted on a vertical axis. The different types of wheels, such as the overshot wheel, used where the fall in a stream is considerable, and in which the water acts both by its weight and by the force of the impact; the undershot wheel, used where there is only a small fall, and in which the water acts largely by impulse; the breast wheel, which in type stands between the two mentioned; and finally the turbines, which are preferred where there is a considerable fall of water, can all be adapted to the running of pumping machinery. The efficiencies of these motors are high, particularly in the case of overshot wheels and turbines, in which they are often as much as 75 per cent.

Hydraulic motors, in which the water is used in driving a piston in a cylinder, much in the same way as steam or compressed air operates, are not used for the pumping of water for country houses, though in the operation of large water works such engines are occasionally met with.

Gas Engines. — Among fuel or internal combustion engines, adapted for pumping water, the first to be mentioned is the gas engine. In this machine the motive force is produced by the expansion of a mixture of gas with air. This, when ignited, explodes, and the force

generated is utilized in moving a piston, which in turn
operates the pump.

The essential requirements for a gas engine (Fig. 18)
are a supply of illuminating gas, or else a special gas
generating plant, and also a supply of water which is
necessary to constantly cool the engine by means of a
water jacket. One of the
advantages of gas engines is

FIG. 18. — GAS ENGINE AND PUMP DIRECT-CONNECTED.

that they can be started at full speed almost as soon as the
gas is lighted. Once started, they require but very little
attendance. There is no fire to be fed, and no nuisance
from smoke and ashes. Gas engines are therefore very
economical machines for the pumping of water. The
drawbacks are the disagreeable odors arising from the
burning of gas, or from small leaks, and the noise of the
exhaust.

Gas pumping engines for domestic use are built in

sizes ranging from 0.5 to 16 horsepower. The gas engine has been developed considerably in other directions during recent years, and large gas engines, developing 1000 and more horsepower, are being successfully built and installed in place of steam engines. They use for fuel the waste gases from coke furnaces or else require special producer gas plants. Gas engines of the form shown in Fig. 18 have been frequently used in pumping water for isolated houses, but in recent years they are being superseded more and more by oil engines.

FIG. 19. — OIL ENGINE AND PUMP CONNECTED BY GEARING.

Oil Engines. — Oil engines for pumping water are similar to gas engines, except that they do not require a gas supply or a gas-producing plant, for oil engines use an independent fuel, such as crude oil or kerosene, which is readily supplied and brought to the engine. The oil is vaporized by heat, and the resulting vapor is ignited either by an electric spark or by other means. The resulting compression of air furnishes the motive power.

The oil engine resembles the gas engine in so far as it requires no boiler and no skilled engineer to run it (Fig. 19). It is simple in construction, reliable and efficient in action, durable in service, and economical to run. It costs but little to operate, can be run intermittently, and is quickly available in case of emergency. Oil engines are built in sizes from 1 to about 20 horsepower; the smaller sizes are more efficient than steam pumps, and compared with gas engines they are somewhat cheaper to run. No danger is connected with their use, and any ordinary intelligent laborer can start and run them.

Great progress has been made during recent years in the design of oil engines. The oil engine is one of the more modern kinds of motors for the raising of water; it is a favorite, particularly in portable form, and in localities where the water is so hard as to be unsuitable for use in steam boilers. Its only drawbacks are the smell of the oil and the noise in operation.

Gasoline Pumping Engines.— In gasoline pumping engines, power is obtained from the explosion of a mixture of air and the vapor of gasoline. They resemble in many respects the gas engines, but the latter and the oil engines are somewhat safer, for gasoline is a highly volatile liquid and must be handled and stored with great care. It is well known that from the fire underwriters' point of view, all mechanical appliances using the vapor of gasoline are considered to be highly dangerous.

Hot-Air Pumping Engines. — Another type of engine for pumping water, which has been used extensively for many years, and which has proved quite successful and economical in use, is the hot-air or caloric pumping engine,

FIG. 20. — RIDER HOT-AIR ENGINE AND PUMP.

two forms of which are in general use, viz., the Rider and the Ericson engines, the former illustrated in Fig. 20. The operating force of the engine is derived from the expansion of atmospheric air in a cylinder by the heat of combustion. The caloric engines are manufactured in various forms, and the fuel which they burn is either coal, coke, wood, gas, oil or gasoline.

One of the reasons for their having proved successful is the fact that they can be run and attended by unskilled persons, and that, after being started, they require very little further skilled supervision beyond an occasional oiling of the working parts. They cannot, however, as a rule, be left for any length of time without occasional attendance, and in this respect they are not as satisfactory as gas, oil, or gasoline engines.

No danger is attached to their operation. Hot-air

engines require a supply of circulating water for the cooling
of the air cylinder. Compared with oil engines, they have
the slight drawback that they cannot be so quickly stopped
when running.

Hot-air engines are built in several sizes, with capacities
ranging from 2 to 25 gallons per minute. They force
water up to a height of from 50 to 150 feet. The power
developed is small, ranging from $\frac{1}{4}$ to 2 horsepower. The
machines are simple in construction and tolerably free
from noise in action. While they have given excellent
service in numerous cases, they have the disadvantage that
some of the working parts are liable to require frequent
repairs.

Steam Pumps. — Steam pumps are not often used for
the supply of water to country houses, but where steam is
already available, as in the case of large institutions which
are equipped with a steam boiler plant, the use of a steam
pump should always be considered. It is quite usual, in
such cases, to use the well-known forms of duplex direct-
acting non-condensing pumps, and sometimes the compound tandem simplex or duplex pumps. Condensing
steam pumps and high-duty pumping engines are considered only in those exceptional cases where high economy
in operation is expected. Where the power exceeds 15
to 20 horsepower, steam pumps are economical in use.
Among the disadvantages are the heavy first outlay for
the pumping machinery, the constant expenditure for fuel
to generate high-pressure steam, and the continuous
skilled attendance required.

For an isolated pumping plant, a combination of direct-
acting steam pump and vertical boiler is much used, which
commends itself because of the compactness of the appa-

ratus. A low-pressure steam and vacuum pump is also manufactured; it is adapted for use in residences and has been moderately successful.

Air-Lift Pump. —Water has in recent years been pumped by means of compressed air. A form of apparatus usually designated as an "air-lift pump" is much used

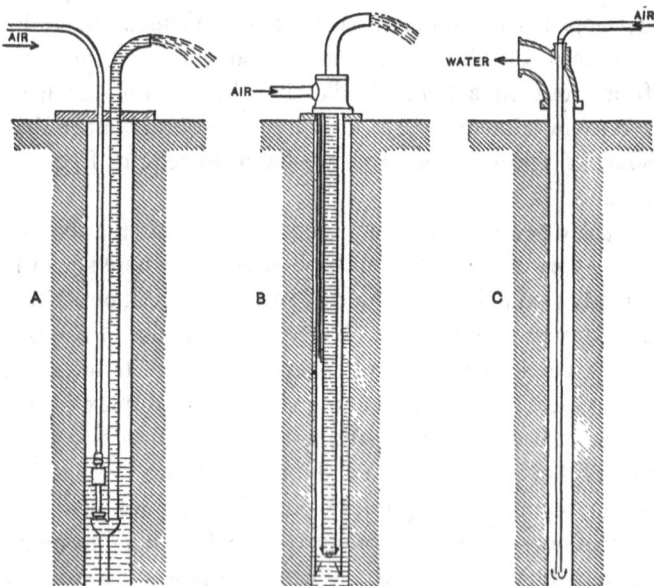

FIG. 21. — THREE FORMS OF AIR-LIFT PUMPS.

because of its simplicity and efficiency, particularly in the case of deep wells. Such air-lift devices afford an opportunity for increasing the yield of wells. From a sanitary point of view, this method of pumping recommends itself because of the beneficial effect derived from the aëration of the water. From the mechanical standpoint, the air-lift device is preferable for deep wells, because it does away with any working parts, such as the pump barrel, which must be lowered into the well below its water line. It dispenses with pumps entirely, and all machinery is above ground. The air compressor required may be run by an oil engine, by an electric motor, a water wheel or turbine, or else by steam.

There are several forms of air-lifting apparatus; in the one shown in Fig. 21 at *A* the air and water pipes are placed alongside of each other in the well casing and connected at the bottom; in another, shown at *B*, a water delivery pipe is placed into the well, and an air pipe passes down through the annular space between the well casing and the delivery pipe; in still another form, shown at *C*, the well pipe is used as a water delivery pipe, and an air pipe is put down into the well, the air being made to escape at the bottom by means of a special device.

The air-lift process of pumping water has been quite successful in recent years, and for the future a more frequent application of the method seems to be assured. Compressed air may also be used either at a constant pressure or else expansively in direct-acting pumps, but this is not a usual form of pumping apparatus.

Electric Pumps. — In recent years, electric house pumps have been used extensively, and this notwithstanding the fact that the use of the electric current is still quite expensive. Electric power for pumping water has become popular and successful because of its many advantages. Electric pumps are compact, extremely convenient and clean in operation; they occupy but little floor space, are generally quite noiseless, create no smell, heat or dirt, and require no handling of fuel or the removal of the waste products of combustion. Electric house pumps can be run without much skilled attendance and require a minimum of care and attention. They can be run intermittently or be fitted up so as to operate automatically. The force operating them is produced by means of a motor, which converts the electric current into power. The motor is either connected directly with the pump or else operates it by means of gearing, belting, noiseless chain drives or worm gears. An incidental advantage of electric pumps is that the power may be transmitted a long distance, and that consequently the pumps may be operated at a distance from the power-generating station or the point of control. In country houses which have an electric lighting plant, the current may be utilized during the day to raise water.

FIG. 22. — QUIMBY ELECTRIC SCREW PUMP.

A popular form of electric house pump is shown in Fig. 22. It consists of a direct-connected screw pump, with motor and pump mounted on the same shaft and bed plate. A belt-driven triplex piston electric house pump, designed to run at a moderate speed, is shown in Fig. 23. Another pump is a triplex electric house pump, operated by worm gearing, the latter rendered noiseless by run-

FIG. 23. — BELT-DRIVEN ELECTRIC TRIPLEX PUMP.

ning in a chamber filled with oil. It consists of three parts, a three-cylinder or triplex pump, an electric motor, and a noiseless worm-gear transmitter. It has no belts, no idler pulleys, no noisy wheel gearing or friction clutches, and is durable and simple in action. The working parts of the transmitter are enclosed in a tight case partly filled with oil, with which the worm gearing is lubricated. The entire outfit is mounted on a single bedplate.

Centrifugal Pumps. — Centrifugal pumps could not, until quite recently, be used to any large extent for the pumping of water, because their power was confined to moderate lifts, not exceeding about 20 feet. A new modification of the centrifugal pump, shown in Fig. 24, differs in this respect, as it is designed particularly for high

FIG. 24. — ELECTRICALLY DRIVEN MULTI-STAGE CENTRIFUGAL PUMP.

lifts, up to several hundred feet. In this new form of centrifugal pump the apparatus is so arranged that the water is pumped in several stages, and the pump is designated as a multi-stage turbine pump. The different impellers are mounted on a single shaft, and the water passes through the impeller chambers in succession; in this way the usually moderate lift is multiplied several times.

Deep-well Pumping Machinery. — Deep-well pumping machinery requires a plunger of suitable size in order to keep the speed of the piston within working limits. This is one reason why deep wells require pipes of larger diameter than is necessary with the ordinary driven well of moderate depth. For deep-well pumping, a single-acting ball valve cylinder pump is the best because of its

simple construction.   Such apparatus (see Fig. 25) is
apt to give considerable trouble when breaks occur in the
working parts which are low-
ered into the well, sometimes
to very great depths.   The
entire machinery requires to
be well-designed and strongly
constructed, because of the
great strains on the working
parts, caused by the stopping
or starting of the machinery.

Pumps in General. — The
above general résumé is sufficient to indicate the points
which should guide one in
the selection of a suitable
apparatus for the pumping
of water.   Whatever the style
or type of pumping engine
selected, it must be borne in
mind that the capacity of
the pump should be such
as to furnish the maximum
daily amount required in a

FIG. 25.—DEEP-WELL PUMP.

few hours, for in the case of country houses, continuous
pumping is generally out of the question.   This statement applies to all fuel pumping engines.   The hydraulic
ram, the various forms of water wheels, and the windmill form an exception to the rule.   In the case of
deep-well pumping machinery the size of the pump
barrel or cylinder must be proportioned to the maximum
yield of the well.

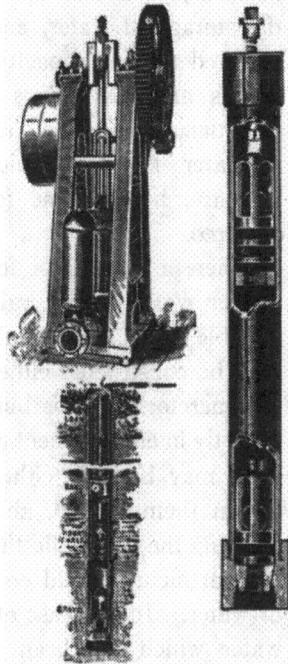

Reservoirs and other Means for Storage of Water. — Having provided the pumping apparatus, it becomes necessary to decide upon the means to be provided for the storage of water, and incidentally, to determine the required storage capacity. In the case of large country houses and institutions an ample storage of water is essential. But even in the case of smaller cottages, a water supply should be provided, flowing under pressure, hence some form of storage tank is always required.

Where a site can be found at an elevation sufficient to deliver water to the building and the grounds under a suitable pressure, a small reservoir may be built. This may be constructed either of earth, of stone masonry, or of concrete; and it is built most economically by putting it partly in embankment and partly above ground. Reservoirs may be built either open or covered; it is best to design them in two, about equal, sections, to permit of cleaning the one while the other remains in service. The compartments should contain at least a ten days' supply of water. In the case of ground waters, and also for all water which has been filtered, covered or arched-over reservoirs, as shown in Fig. 26, are required in order to exclude, in the one case the sunlight, in the other dust and atmospheric impurities. It has been mentioned heretofore that sunlight promotes vegetable growth in the case of ground waters, imparting to them a bad taste and odor.

In country districts, where stones abound, a service reservoir may sometimes be built at a small expense, but as a rule its construction, if built in brickwork or of concrete, and if made water-tight with asphalt or cement,

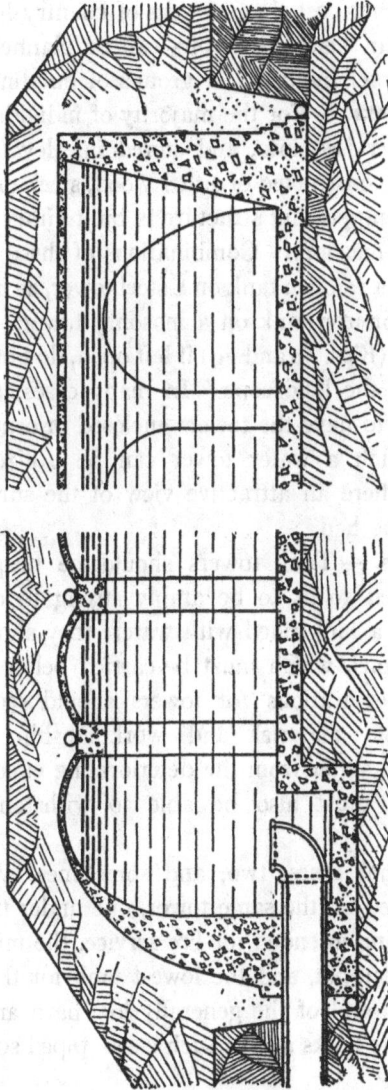

FIG. 26. — COVERED UNDERGROUND RESERVOIR OF CONCRETE MASONRY.

involves a higher cost than owners of country houses are willing to incur, hence reservoirs can be planned only in the case of larger buildings or groups of buildings.

Elevated Tanks. — For the majority of individual buildings, water is raised into, and stored in, elevated tanks. These may be constructed of either wood, steel, or wrought iron, and the supporting structure is built either in wood, in steel, or of masonry. Combinations of these materials occur, such as a wooden tank on a steel tower, as in Fig. 27, or a wooden or iron tank on a masonry tower. Wooden or iron towers (Figs. 28 and 29) if left open, do not enhance the beauty of the landscape. In many cases, therefore, owners prefer to have the towers enclosed or ornamented; quite often such a water tower can be utilized as an observatory where an attractive view of the surrounding country may be had.

Tank Towers. — Tank towers should be proportioned and constructed so as to be amply strong to carry the heavy load of a tank filled with water; they should have good foundations, which must be carried below the frost line. The specifications for towers should require the best obtainable material and workmanship, because exposed towers suffer from the deteriorating effects of the weather; they must also be able to withstand heavy wind pressures.

In exceptional cases, two, and sometimes even three tanks are placed on the same tower, one under the other, the highest being intended for fire service, the middle one for the house supply, and the lowest tank for the supply of the grounds and of the generally low barn and stable buildings. The tanks are, in such cases, piped so as to be interconnected.

FIG. 29. — AN IRON TANK ON A STEEL TOWER.

FIG. 28. — A WOODEN TANK ON A WOODEN TOWER.

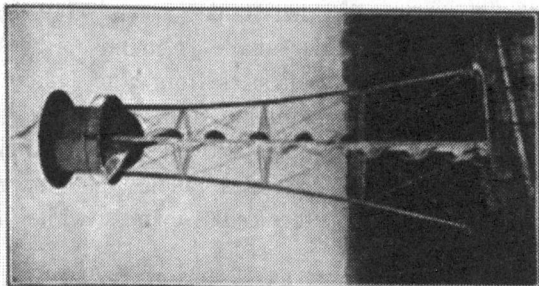

FIG. 27. — A WOODEN TANK ON A STEEL TOWER.

Wooden Tanks. — Wooden tanks are always built round, because the circular shape is found to be the best to secure tightness. All sizes from 3 to 32 feet in diameter, and with staves ranging from 3 to 24 feet in length are manufactured, the respective capacities ranging from 150 to 120,000 United States gallons. The special lumber used for tanks is either cedar, cypress, white pine or Oregon fir.

Cypress lumber is used largely in the western and southwestern states, while preference is given in the eastern states to white pine and cedar. Some tank manufacturers claim that for exposed tanks nothing is better than Red Gulf cypress, but others prefer the pine wood. Cypress wood is soft, coarse-grained, and porous, and easily becomes water-soaked. Tanks of this material leak more and are more easily affected by frost in the northern climate, whereas pine lumber is close-grained and such tanks are more readily made tight and are otherwise just as durable.

Tank lumber should be clear, selected stock, thoroughly air-dried and free from knots, wormholes and shakes. The staves and the bottoms are made of $2\frac{1}{2}$ or 3-inch stock; the staves, which should not be wider than 8 inches, are surfaced and sawed to the proper bevel. They are doweled together by means of dowel pins, placed about 4 feet apart. All staves must be of the full length and should not be spliced.

The strength of wooden tanks depends principally upon the iron or steel hoops with which they are encircled, and which always have lugs and bolts to tighten the tanks. Hoops are either flat or round; the latter, which are required by the underwriters, are preferable because they

can be better examined and protected from rust. The
diameter, number, and the spacing of the hoops depend
upon the size and capacity of the tank. All hoops should
be well painted before being put on and should be given
another coat after the tank is erected. It is essential, in
the construction of a wooden tank, that the weight of water
should be supported entirely from its bottom, and therefore
sleepers are laid under the tank bottom, and are cut to the
circumference of the tank, but a few inches shorter.
Wooden tanks, when left standing empty and exposed to
the sun, soon become leaky; those intended for winter
service should always be frostproofed in the best possible
manner. Wooden tanks of an oblong or square shape
cannot be made to stay tight without being lined with
sheet metal the same as the inside house tanks; for this
reason they are but rarely used.

Iron Tanks. — Iron tanks are made either square or
round, and are built of cast iron, or else of wrought iron
or tank steel. Square cast-iron tanks, consisting of numerous sections bolted together with tightly made joints, are
used only in the interior of buildings. Outside tanks
are made of wrought iron or of steel, with riveted joints;
such tanks are nearly always
round in section. They cost
from 50 to 100 per cent more
than wooden tanks, without
having greatly superior advantages to offset the larger cost.

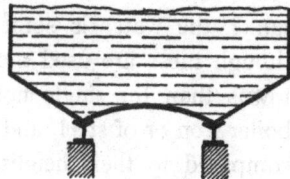

Iron tanks are commonly
made with a flat bottom, but

FIG. 30. — IMPROVED FORM OF
IRON TANK BOTTOM.

this requires the placing of a number of iron beams at
short intervals under the tank bottom. A considerable

saving can be effected in the cost of the supports by making the tank bottom either hemispherical, as in Fig. 29, or else by using a compound shape and supporting the tank directly and only at its circumference as in Fig. 30. An incidental advantage of such more modern forms of tanks is that the bottom can be better inspected and repaired.

Wooden versus Iron Tanks. — The reasons why wooden tanks are more often used than steel tanks are that the latter are more difficult to erect, that they give trouble by reason of sweating, and that they soon rust, if the outside paint is not constantly renewed. Iron tanks are also more difficult to protect against freezing. But where very large storage capacities, exceeding 120,000 gallons, are necessary, iron tanks must be used. Hence we find tanks, such as those shown in Fig. 29 and Fig. 30, largely used in the water supply of institutions, villages and small towns. Such large sizes, however, are but rarely called for for country mansions, and therefore wooden tanks are preferred for these. In practice, it is found that the latter last from 15 to 25 years, are not so liable to freeze, do not rust, keep the water pure and clear, and can be bought and put up for much less money.

Standpipes. — Instead of providing elevated storage tanks, the water is sometimes pumped into standpipes, though these are used more for the supply of villages or towns than for single houses. They are built of either boiler iron or of steel, and are always small in diameter as compared to their height. Such standpipes are erected directly on the ground or on suitable stone foundations, and the water which is pumped up into the lower portion forms the support for the water that is stored for use under the available or required pressure. Standpipes are often left unenclosed and give considerable trouble in winter time by the forming of ice. It is safer and better to enclose them with masonry as a protection against freezing, but this adds greatly to their cost, and hence standpipes are not used to any extent in the case of country buildings.

Inside or House Tanks. — The water required for a dwelling may be stored in inside tanks, which may be located in the attic, or in the case of institutions sometimes placed at the top of the main staircase tower. Such tanks are built of wood lined with copper, of cast iron, of slate, and finally of wrought iron or steel. They are usually square or oblong in shape, but can be made to fit a space of irregular dimensions and shape. The objection to inside tanks is that their size must necessarily be limited, not only owing to want of space, but also because of the heavy weight of water, which cannot always be readily carried, hence they provide but an insufficient storage of water. Where an outside tank can be provided, it is much better to do so.

Pressure Tanks. — Another method of storing water consists in providing pressure tanks. These are closed riveted boiler iron or steel tanks, of any required dimensions, conveniently placed either on or, more often, below the level of the ground. Owing to this location, they do not require any expensive supports such as the elevated tanks have, though their weight necessarily requires good foundations. In some systems the pressure tanks are designed to be kept two-thirds full of water, the remaining one-third being filled with air under pressure.

Pressure tanks are always cylindrical in shape and may be placed either horizontally or vertically. When used without an air compressor, pressure tanks have the disadvantage that the flow of water becomes reduced as the water level is lowered in them.

Some of the advantages of such tanks are that the pipes and tanks can more readily be made frostproof; that the water is kept at a cooler temperature in summer, and that it is not subject to pollution during storage, because the tanks are closed.

FIG. 31. — A SIMPLE PRESSURE TANK SYSTEM.

In using pressure tanks, it is necessary either to provide a high initial pressure, so that the last few gallons stored may be delivered under a sufficient pressure, or else to renew pumping before the tank is empty. For use in country houses several systems with pressure tanks, operated either with hand or power pumps, have come into use in recent years. A simple system for use in a small country house, with vertical tank and handpump, is shown in Fig. 31, but the same appliances, of larger dimensions, may be used in connection with any of the pumping apparatus heretofore described.

The Acme Water Supply and Storage System. — A good modification of the pressure-tank system, consisting of pump, air compressor, water tank and air tank, was invented and patented years ago by a well-known engineer, the late W. E. Worthen. Many examples of this system, known in its improved form as the "Acme Water Supply and Storage System," have been constructed, not alone for country towns, but also for isolated dwellings. The system has many sanitary and constructive advantages, provides an excellent fire protection, and does away with the use of fire engines. The water stored in the underground tanks is kept cool, remains pure, and cannot become contaminated as in the open tanks. Moreover, in the case of underground sources of supply, there is no exposure of the water to the sun's rays, hence there will be no annoying growth of algæ. There is also no trouble from the freezing of the water in the reservoir.

The chief advantages gained by using, as is done in this system, an additional tank for compressed air are additional water storage capacity, the water tank being com-. pletely filled with water, and the fact that the last gallon of water in the tank is supplied under the initial pressure. This cannot be accomplished in those pressure-tank systems which do not have an air tank. For the supply of buildings composing a military post, this system has the advantage, from the strategic point of view, that

there is no standpipe or elevated tank at which the
guns of an enemy could be aimed, hence the water supply
cannot be cut off or destroyed in case of war. In all
cases where objection is raised to the appearance of an
elevated tank, or to the heavy additional expense required
to make it look well, this system is well adapted and worth

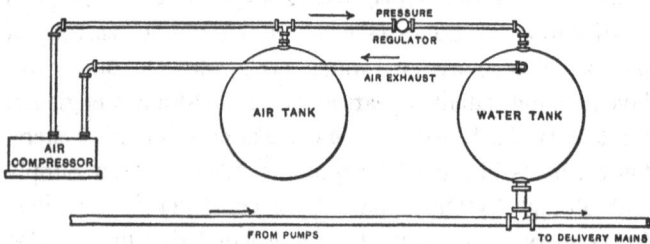

FIG. 32. — DIAGRAM ILLUSTRATING THE " ACME " PRESSURE
TANK SYSTEM.

investigating. The chief features of this system are
shown in Figs. 32 and 33; and in the illustrated examples
given at the end of this part of the book, several installations of the system for country houses are described more
in detail.

Water Distribution. — Having pumped the water and
stored it in elevated tanks, or in reservoirs, or in pressure
tanks, we have to provide the main conduit pipe to bring
it to the places of consumption. In the case of important
buildings, it is advisable to install this main conduit in
duplicate, so as to provide against the complete cutting
off of the supply in case of a break. Larger conduits are
always built of cast-iron pipes with lead caulked joints, and
when the pressure is heavy, considerable attention must
be given to the proper jointing of the pipes to avoid constant leakage and consequent waste of water. Conduits

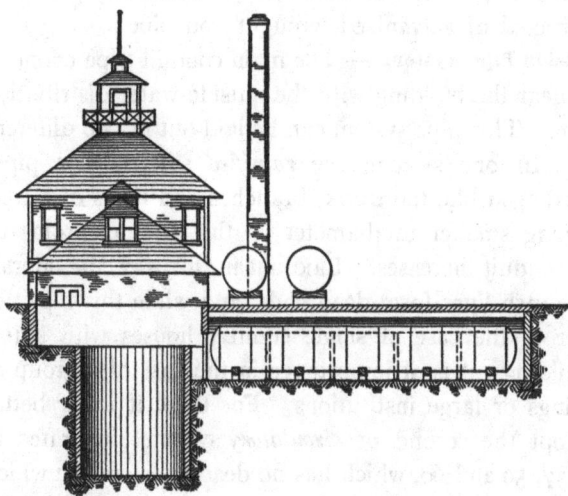

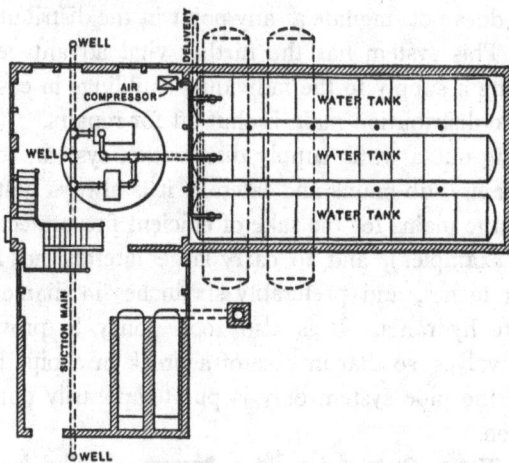

Fig. 33. — "Acme" Water Supply System with Water and Compressed Air Tanks.

which are 4 inches in diameter or smaller are usually
constructed of galvanized wrought-iron pipe.

Outside Pipe System. — The main conduit pipe connects
at or near the building with the outside water distribution
system. This pipe system can be laid out in two different
ways. In one system, the *ramified* system, the pipes
are arranged like the trunk, branches and twigs of a tree,
becoming smaller in diameter as the distance from the
main conduit increases. Laid out in this way, the laterals
and branch lines form dead ends, and while this may not
matter in the case of single country houses with a few
outbuildings, it is of importance in the case of a group of
buildings or large institutions. For these it is far better
to adopt the second or *circulatory* system, illustrated in
Figs. 37, 59 and 60, which has no dead ends, and in which
the extreme pipe ends are looped so that the water circulates and does not stagnate at any point in the distribution
system. This system has the further vital advantage of
maintaining a supply to the individual buildings in case a
part of the distribution main is shut off for repairs.

In laying out a water-supply distribution system, comprising mains, sub-mains and laterals, it is always best to
provide large mains for the sake of efficient fire protection
(see the examples), and to carry large laterals, not less
than four inches, and preferably six inches in diameter,
to the fire hydrants. It is also good policy to provide
plenty of valves, so that in case of a break in a pipe line
a part of the pipe system only is put temporarily out of
commission.

Inside Water Distribution Pipe System. — The inside
water-service system is, even in the case of a single country
mansion, more intricate than the outside water supply;

in the case of a group of buildings the work of planning the inside distribution becomes at times quite complicated. There is always required a double system of water pipes, one for cold and the other for hot water. Sometimes a third system, for artificially cooled water, is added; and in other cases country residences require a double pipe system, for rain and for well or spring water. In its essential details the inside water service of country houses does not differ much from that of city houses.

The inside distribution system is either a so-called *header* system, in which separate runs are provided for each riser, all coming from a single center, manifold, or header, or else it is a *circuit* system, in which large mains for cold and for hot water are provided, from which short branches are thrown off to the rising lines.

The first system requires more piping, but the pipes become of a smaller size and the control of the lines, valves, and drips is placed at one centrally convenient point. It is more expensive in first cost, but it is handier to look after. In the second system, the valves controlling the risers are scattered, therefore it is not so convenient and compact for the engineer in charge of the plant, but it has some advantages; it is cheaper, and does not require so many lines of piping and consequently not so much space at the cellar ceiling.

General Arrangement of Water Pipes. — In arranging the water piping in the interior of buildings, several essential rules must be followed, such as the placing of pipe lines where they will not freeze, or, where they are unavoidably exposed, the suitable protecting of the pipes against damage by frost. Noise and water hammer in the supply pipes must be prevented as much as possible and the heating up of a cold water line from an adjoining hot water pipe, steam line or flue, should be guarded against. In laying out a supply system, the aim should also be to arrange the pipes, and proportion their sizes, so that there will be a good, constant and uninterrupted flow at the faucets in a building, if many of them are kept running simultaneously.

Hot Water Supply. — In the hot water supply system, circulation pipes should always be provided to permit of the instant drawing of hot water at any faucet in the house. This measure is a good help in preventing much useless waste of water. In running a circulation pipe it should be remembered that if it rises in loop form above the highest fixture, an air cock should be provided at the top of the loop to avoid air binding.

Material for Supply Pipes. — Regarding the materials used for the water-supply service, it is customary to run these lines with galvanized or asphalted wrought-iron pipe, with screw joints, where the outside mains and laterals are not larger than four inches. The outside hydrants are either post hydrants or flush hydrants, and it is well to provide a shut-off gate valve at each of them. In the case of institutions, all hydrants should be post hydrants with indicators. On long lines it is advisable to provide blow-offs at low points, so that the line may be emptied; this is essential where a pipe line is for summer use only. At high points in a main conduit line, air valves are provided to prevent air binding, and these are sometimes arranged to work automatically.

Water Supply for Fire Protection. — Of special importance in the case of country residences and mansions is the matter of their *fire protection*. As a rule such residences are much more in danger from fire than city houses; they are usually located at a distance from any regular fire department, their supply of water often falls short of the demand, quite usually they are built in a flimsy manner, with little or no regard to safety from fire; and moreover, they are but seldom provided with proper means for the extinguishment of a conflagration.

It is quite exceptional to find proper attention paid to the matter of fire protection, though in some instances money is uselessly spent on worthless fire apparatus, put in largely with a view of obtaining a reduction in the insurance rates. While it is true that a rebate in the insurance rates is given by underwriters, this only applies — and properly so — where a fire protection system is well arranged, and put in in accordance with their requirements.

The means for *fire prevention* do not come within the scope of this book, hence cannot be discussed; those for *fire protection* comprise both indoor and outdoor water fittings and appliances, which should be briefly mentioned.

Many houses in the country are provided with an inside fire standpipe, too small in size to be of any use in an emergency; the water pressure is often insufficient, and the fire hose installed is of a worthless type, while outdoor fire appliances are lacking entirely in the case of many fine country estates.

Outside Fire Hydrants. — Every mansion, and every institution located in the country, should have a sufficient number of outside fire hydrants. In determining the distance between the hydrants, it is well to remember that considering the life of both, cast iron pipe is cheaper than hose. The mains supplying the hydrants should be ample in size. For use at the hydrants, there should be provided best-quality large rubber-lined hose, hose spanners, hose carts and the best form of fire nozzles. The entire fire-fighting apparatus should be constantly kept in good working order, for otherwise it may be found to be unserviceable just when most wanted.

Not long ago the services of the writer were engaged to lay out the water supply of a country estate, comprising a fine mansion, a large stable, and a gardener's lodge. With a view of providing some protection against fire, he specified a large water supply main, with a number of outside fire hydrants, located near the buildings. But when the cost of the proposed work was communicated to the owner, who was a rich man and could well afford the expenditure, he called the writer a "fool" for making the suggestion, and the work was not carried out as planned. Some day, this fine unprotected mansion, which is located several miles away from the nearest village fire engine house, may burn to the ground, and when that day comes the owner will have every reason to apply the above epithet to himself.

Fire hydrants and fire hose are useless unless an adequate storage of water under a suitable fire pressure is maintained. Where this is provided, it takes the place of the steam or hand fire engines of a fire department.

Portable Fire Engine. — In the case of large estates which have a natural or artificial lake or pond within the grounds, it may be practicable to keep on hand a portable steam or hand fire engine, with sufficient and ample size fire hose, but this should only be regarded in the light of "auxiliary" fire protection. It is always better to install a fire protection equipment which is instantly available in case of need; with this in view, all fire apparatus should be kept in working order.

Inside Fire Standpipes. — Regarding the inside fire protection, it should be pointed out that in two- or three-story houses, having a house tank placed in the attic, a fire standpipe, supplied from the tank, has an entirely insufficient pressure. Fire standpipes should therefore be connected with the outside main, and be supplied directly from the main reservoir, the elevated tank or the pressure tank.

In the house, there should be outlets on the standpipe with fire valves and fire hose on each floor. Plug cocks are not recommended for fire valves, because after being installed for some time they set hard and are then very difficult to open. The highest quality linen fire hose should be purchased, the unlined hose being preferable for inside use, as it is less liable to deterioration than rubber-lined hose. In addition to the fire hose, a country house should have a number of fire pails kept constantly filled, besides a few portable pneumatic or chemical fire extinguishers. The use of the common hand-grenades is to be discouraged.

EXAMPLES OF WATER SUPPLY SYSTEMS.

In the following I give a few plans and descriptions of the proper arrangement of the water supply for country buildings, some of them being taken from my own practice.

I. An example of a system of water supply for a country residence, which includes provision for fire protection, is shown in plan in Fig. 34. The water supply for the house and grounds is obtained from a brook which runs at all times a fairly large stream of water. The water is pumped to an outside storage tank by means of two Rife hydraulic rams or

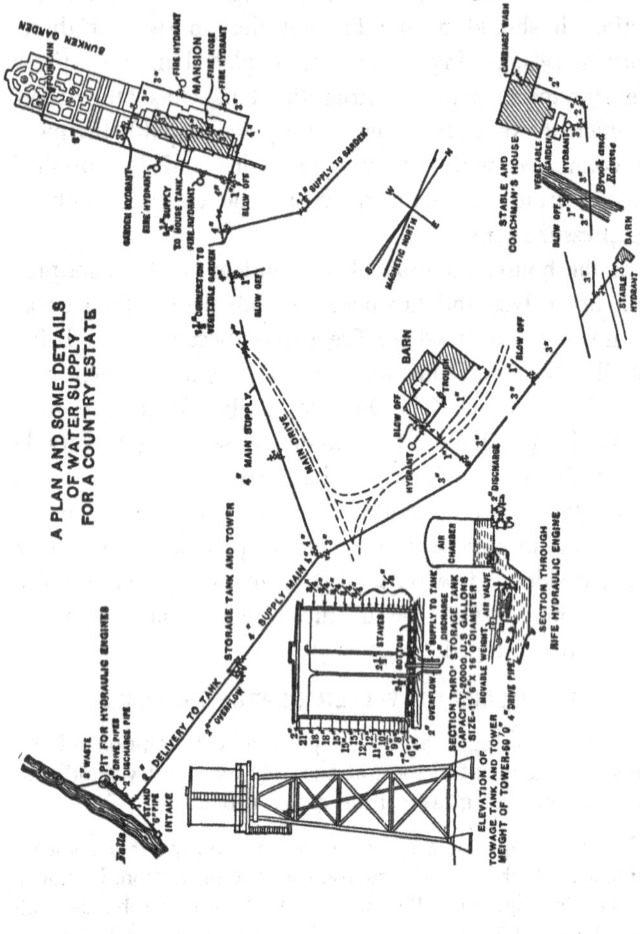

Fig. 34. — A Water Supply System from Two Hydraulic Rams.

engines, operated by two separate 4-inch drive pipes, and having one common 2-inch discharge pipe. The elevated tank is a round one, built of wood, and has a capacity of 20,000 gallons. It is supported on an open wooden tower, the height of which is approximately 60 feet. The falling main from the tank is a 4-inch pipe, and the main supply conduit is run, 4 inches in size, to the mansion. A distribution system, consisting of 3-inch pipes with 3-inch branches to four fire hydrants, is provided all around the house. The mains are purposely made large because the available tank pressure is not very great, the ground where the tank tower is erected being only a few feet higher than the site for the mansion.

An inside fire standpipe is also supplied from the outside service under direct pressure, while the house supply for domestic use is obtained from an attic tank fed by a branch from the 4-inch main. A 3-inch pipe line is branched off from the main and is carried past a sheep barn, with one branch for a watering trough and another branch at a second barn for farm horses. The 3-inch line is then continued to the main stable and carriage house, where another fire hydrant is provided in addition to the supply for the coachman's rooms and the carriage wash. With a $2\frac{1}{2}$-inch fire hose attached to the fire hydrant, a fire stream was thrown over the house under the available pressure, which was approximately 32 pounds on the ground level.

II. As an example of a successful and exceptionally complete water supply and fire protection of a large hotel building in the country, a brief description is given of the water supply of an American hotel, "The Mount Washington," situated in the White Mountains, near the base of the chain of mountains known as the "Presidential Range." This large building, shown in two views in Figs. 35 and 39, was provided by the author with two distinct systems of supply, namely, one a pumping supply from a series of wells, and the other a gravity supply from a mountain brook.

The wells are 17 in number (see Fig. 6). They were obtained by means of well-driving machinery, an 8-inch casing having been first driven down through the gravel to a depth varying from 30 to 40 feet. Into this casing a 5-inch galvanized iron well pipe was lowered; each well was provided with a cleanout and with a valve to shut it off, and the different wells were located on both sides of a main suction line, which was made 5, 6, 8, and 10 inches in diameter, and which delivered the water to a closed suction chamber connected with the pumps. The aggregate yield of the seventeen wells was about

FIG. 35. — VIEW OF MOUNTAIN HOTEL.

1050 gallons per minute; the water was pure in quality and of a very low temperature.

The pumping station contains two electric triplex double-acting direct-connected pumps (see Fig. 36), namely a house pump rated at 350 gallons per minute, and a fire pump rated at about 800 gallons per minute. The illustration shows the pumps as set before the pump house was built around them. These pumps deliver the water through an 8-inch pumping main into a series of three wooden tanks, each of 50,000 gallons capacity, located on a plateau on a hillside opposite the hotel. From the tanks a 12-inch delivery pipe was carried to the site of the hotel. The elevation of these tanks is about 150 feet above the basement of the hotel, thus providing a pressure of about 66 pounds per square inch.

The 12-inch main running to the hotel is sub-divided, as shown in Fig. 37, into two 8-inch lines, running along the two fronts of the building. These lines are interconnected at the ends and also at intermediate points so as to form a continuous loop. From the 8-inch main about twenty 6-inch laterals are run to the outside fire hydrants, while there are also a number of 3 and 4-inch supplies for the inside plumbing, the elevator service and the refrigerating plant.

FIG. 36. — VIEW OF THE TWO ELECTRICALLY OPERATED PUMPS.

In addition to the well supply, a gravity supply is brought a distance of over four miles through an 8-inch cast-iron pipe line which connects with the 8-inch pumping from the pumping station. The mountain reservoir contains approximately one million gallons of water, which are stored there by means of a plain wooden dam thrown across the stream (see Fig. 38). The fire pump is arranged so as to draw either from the wells, or, by means of a separate suction line, from an artificial nearby lake. With the pressure from the tanks on the mountain several efficient fire streams were thrown from the hydrants reaching to some height above the top of the building (see

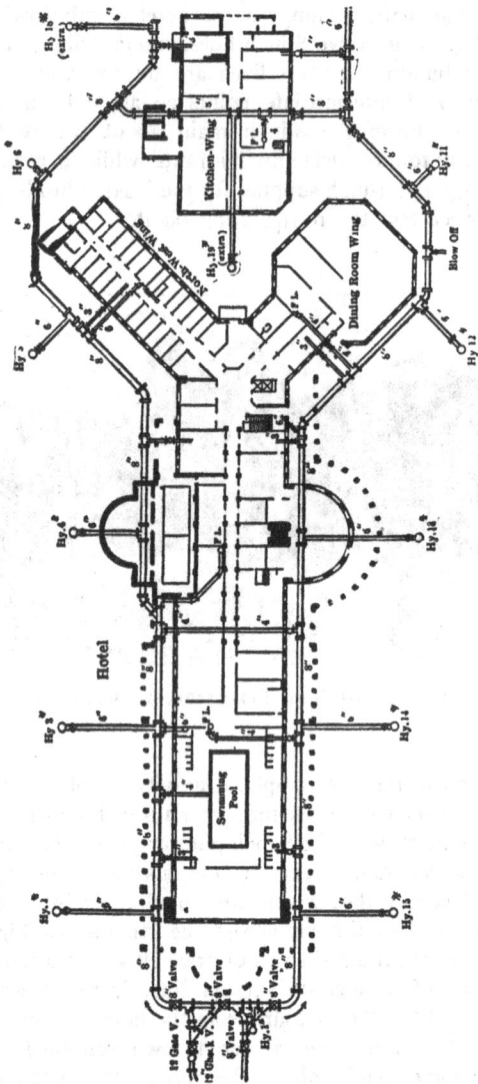

FIG. 37. — PLAN SHOWING WATER MAINS OF HOTEL.

FIG. 38. — RESERVOIR AND DAM WITH GRAVITY SUPPLY MAIN.

Fig. 39). In case of an accident to the 12-inch main the water can be delivered from the tanks to the hotel through the 8-inch pipe line, and vice versa.

FIG. 39. — VIEW OF FIRE STREAM FROM HYDRANT.

III. The country estate illustrated in the accompanying plan (Fig. 40) was recently fitted up by the writer with a modern system of water supply, with a pressure tank installation, intended to furnish an abundant supply not only for the needs of the mansion, stable, cottage and farm buildings, but also for lawn sprinkling, for watering the vegetable garden and for fire protection.

A preliminary examination and survey of the estate and its immediate surroundings had shown that three methods were available for the solution of the problem. In all of these the supply was to be obtained by pumping from a small lake, protected against pollution by strict State laws.

A solution of the problem of water supply which naturally suggested itself first, was the use of an elevated water storage tank, to be erected on a supporting tower on the grounds near the mansion, and to be supplied with water from an engine-driven pump. A windmill pump, which would have been both simple and economical, was found to be out of the question in the particular locality considered, on account of the uncertainty of sufficient wind pressure during the summer time. The pumping plant, therefore, located in a small pump house to be built near the shore of the lake, was to comprise a triplex power pump directly geared to a gasoline or an oil engine. This

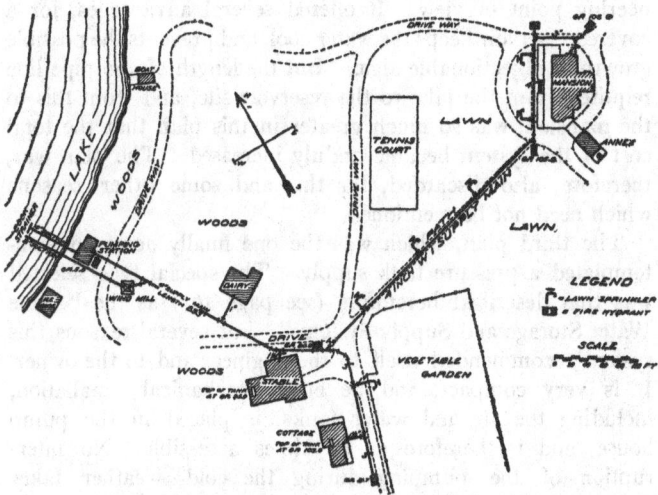

FIG. 40. — PLAN OF A PRESSURE TANK WATER SUPPLY SYSTEM.

project, however, was rejected because of the undesirability of a tank tower being placed close to the residence. No other location for the tower could be found on the owner's property, and while it was possible to render the tower attractive by architectural ornamentation, this would have added considerably to its cost.

In a second plan, the construction of the pump house and of the pumping plant remained practically the same, but the substitution of an underground reservoir for the unsightly tank tower was proposed. On a neighboring property, south of the estate, about 2000 feet away, a hillside was found at an elevation of about 100 ft. above the knoll where the mansion stood. This project involved the obtaining of permission to build the reservoir on the site described, also the obtaining of the right of way to lay the pipes, either by purchase or by other agreement. Such an underground reservoir would have been a very satisfactory solution of the problem from a purely engineering point of view. It offered several advantages, for a covered reservoir keeps the water cool, and prevents the possible growth of objectionable algæ. But the length of the pipe line required from the lake to the reservoir site, and from this to the mansion, was so much greater in this plan that the total cost of the system became unduly increased. The plan was, therefore, also discarded, for this and some other reasons which need not be mentioned.

The third plan, which was the one finally accepted, contemplated a pressure tank supply. The special type selected was that described heretofore (see page 161) as the "Acme Water Storage and Supply System." For several reasons this system recommended itself to the engineer and to the owner. It is very compact, and the entire mechanical installation, including the air and water tanks, is placed in the pump house, and is therefore at all times accessible. No interruption of the pumping during the cold weather takes place, provided the pump house is kept warm. Then again the water can be forced out in this system under any desired pressure, which constitutes a distinct advantage over a system with either an elevated storage tank or with underground supply reservoir. The advantage named is of some importance as a fire protection measure. But the chief advantages of the third plan were that the storage of water was effected on the owner's grounds and not on some neighboring higher property; the distribution pipe lines became shorter; the water was kept much cooler in the water tank, from which air and light are excluded, and the air, coming into intimate contact with the water, had a purifying influence upon the character of the supply.

In a pressure tank system of this kind two air-tight steel or iron tanks are installed, one being the water tank, the other the air storage tank. The pumping machinery forces water into the bottom of the water tank, while an air compressor forces air under pressure into the top of the air tank. Both tanks are connected at the top by an air pipe, and thus air under suitable pressure acts upon the water which it discharges through the bottom outlet of the water tank, which in turn connects with the force or supply main.

A view of the small pumping station, taken from the lake, is shown in Fig. 41. It is a one-story brick structure, with a pump

FIG. 41. — VIEW OF THE PUMPING STATION AT THE LAKE.

room which is 12 by 20 feet in size and which contains the engine, the pump and the air compressor. In a wing, which is about 15 feet square, the two tanks are placed. The plan (Fig. 42) of the pumping station shows the entire arrangement very clearly, and the two interior views which follow help to explain it. A staircase leads from the entrance to the pump-room floor, which is about 5 feet below the grade level. Such a location, partly underground, helps in keeping the building fairly warm during mild winter weather. A smoke flue is provided so that if desired a heating stove may be set up if

the system is wanted in winter time, for instance to provide
fire protection to the main buildings of the estate.

The machinery of the pumping station consists of a triplex
plunger pump, having a capacity of 75 gallons per minute,
a gasoline engine of 10 horsepower, and an air compressor

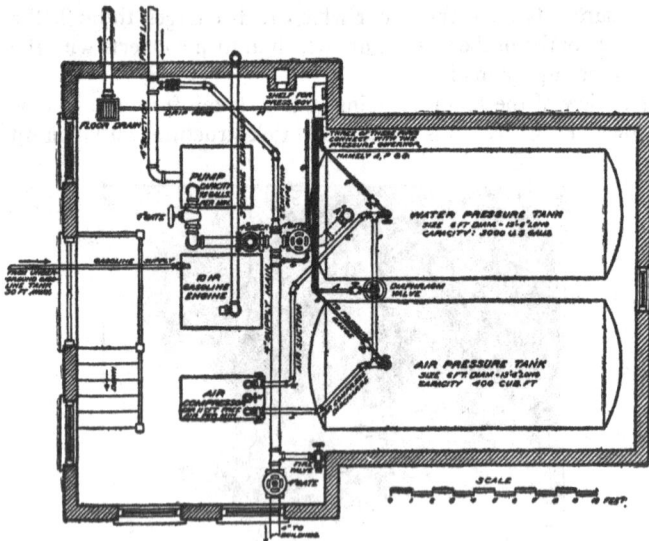

FIG. 42. — PLAN OF PUMPING STATION SHOWING PUMPING
MACHINERY, AND THE WATER AND AIR TANKS.

having a capacity of 11 cubic feet of free air per minute (see
Figs. 43 and 44). The pump and the compressor, shown
in the photographic view, Fig. 43, are mounted on the same
shaft, which is provided with friction clutches, enabling the
throwing out of operation of either the pump or the compressor.
Fig. 45 shows a diagrammatic section through the pump
house, with the water and air tanks and all their pipe connections.

The size and dimensions of the water tank were fixed by
determining the desired amount of storage, by considering the

FIG. 43. — INTERIOR VIEW OF PUMP HOUSE, SHOWING ENGINE, PUMP AND AIR COMPRESSOR.

FIG. 44. — ANOTHER INTERIOR VIEW OF PUMP HOUSE.

probable daily domestic consumption, the volume needed in case of fire, and the intervals required between pumping.

The size of the air tank depends chiefly upon the air pressure to be carried. No matter what the pressure may be in the air tank, the same pressure is always carried in the water tank. The air tank should have enough air under suitable pressure so that all the water stored may be delivered under the given water pressure. If the water and air tanks are to be of the same size, the initial air pressure carried in the air tank must be double the water pressure. Water and air are always pumped simultaneously, and the air compressor exhausts during pumping the air from the top of the water tank and transfers it to the air tank. In Fig. 45 the air suction pipe is marked E, while the air discharge pipe is marked I. Another pipe, M, controlled by a valve, is the free air suction and is used only in case some air has been lost by leakage. It should be noted that in this system the pressure carried in the air tank is always greater than that in the water tank, except when the last gallon of water is discharged, at which moment both pressures are equal.

Fig. 45 also illustrates the apparatus intended to control the air pressure on the water tank, and this is shown, on a still larger scale, in Fig. 46. It consists of a diaphragm valve on the air pipe (see view, Fig. 45), controlled and operated automatically by a governor. By adding or removing weights attached to the lever shown the water pressure carried may be increased or decreased within certain limits.

A water storage of 3000 gallons was considered to be ample. It was stipulated that one fire hydrant with 50 feet of 2-inch fire hose and with 1-inch fire nozzle should discharge water for about 40 minutes under a sufficient pressure to reach over the roof of the house. A fire nozzle under these conditions, with a pressure of 27 pounds at the hydrant, discharges per minute 75 United States gallons, or 3000 gallons in 40 minutes. The daily domestic consumption was estimated to be below 3000 gallons. The water tank was therefore made 6 feet in diameter and $13\frac{1}{2}$ feet long, and the air tank was made of the same size. The pressure of water at the pump house had to be

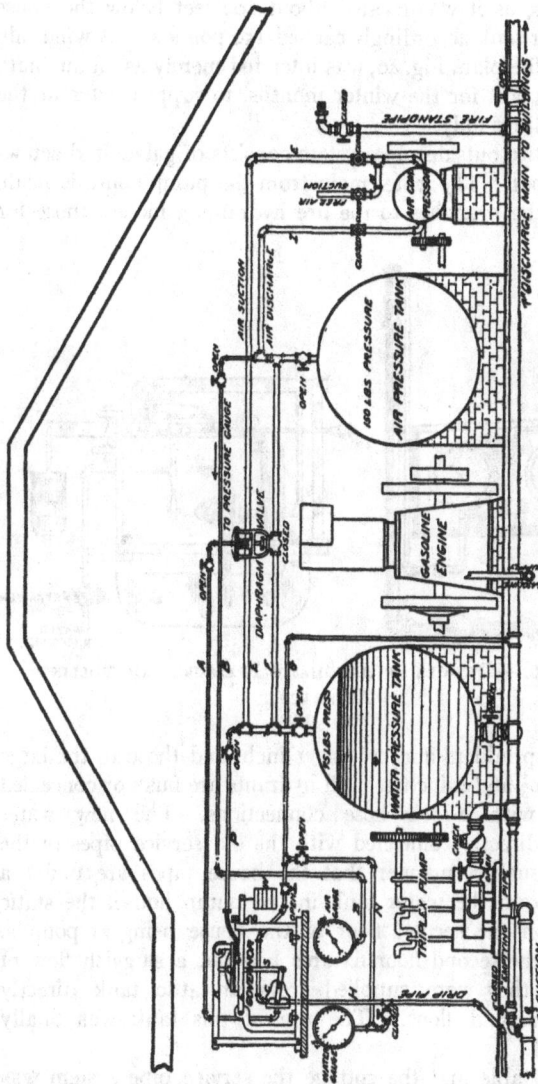

FIG. 45.— DIAGRAMMATIC SECTION THROUGH THE PUMP HOUSE, TO ILLUSTRATE WORKING PARTS OF THE "ACME" PRESSURE TANK SYSTEM.

75 pounds, as it was located about 100 feet below the house and the air tank accordingly carried 150 pounds. A windmill, shown in the plan, Fig. 40, was intended merely as an auxiliary pumping plant for the winter months, to supply water to the farm buildings only.

The entire outside pipe system consists of galvanized screw-jointed pipes. The water main from the pump house is made 4 inches, the branches to the fire hydrants 3 inches, those for

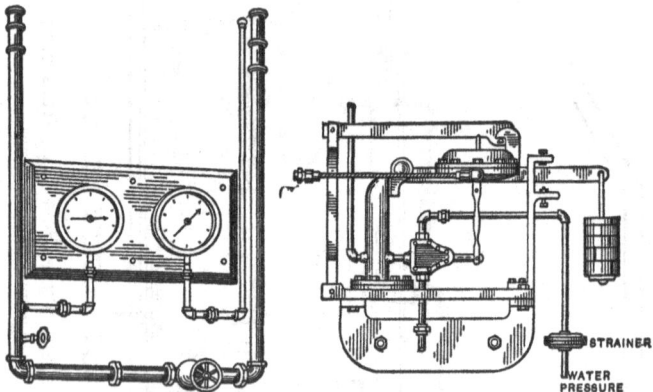

FIG. 46. — DETAILS OF AUTOMATIC PRESSURE CONTROLLING APPARATUS.

the lawn sprinklers and cottage 1 inch, and those to the large buildings 1½ and 2 inches. All hydrants are flush or concealed hydrants with 2-inch hose connections. The new water system is directly connected with the old service pipes of the house in such a manner that the house pipes are under a pressure from the water tank in the pump house, the static pressure on the second floor of the house being 21 pounds. Formerly the second floor fixtures had but a sluggish flow of water, as they were supplied from an attic tank directly over the second floor. The use of this tank was finally abandoned.

In the stable and the cottage the service pipe system was

arranged somewhat differently, for during the winter months these were to be supplied from the windmill pump, hence attic tanks were installed in both buildings, but the connections were so made that during the summer both buildings are supplied from the direct-pressure system.

The operating expenses of the plant were approximately as follows: — with gasoline at about 12 cents per gallon, the engine required about 1¼ gallons per hour, or an expenditure of ¼ cent per minute, hence to pump 3000 gallons of water cost about 10 cents, not including the cost of attendance, which was but slight, and the cost for oiling the machinery.

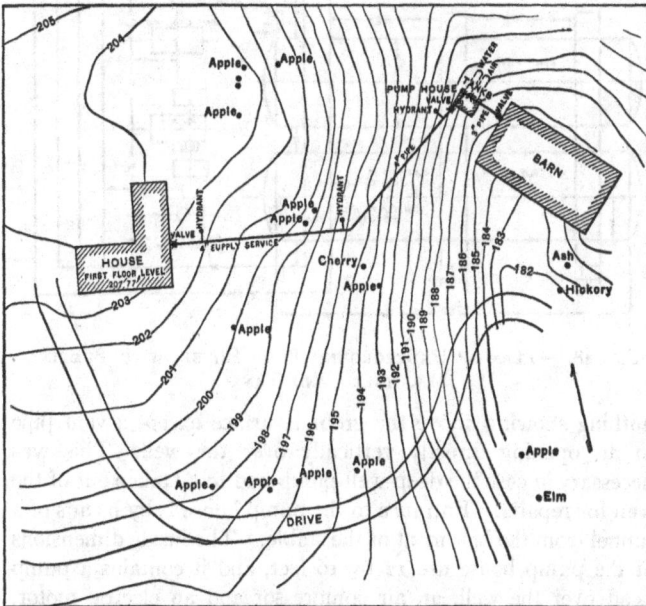

FIG. 47.— PLAN OF WATER SUPPLY BY PRESSURE TANKS FOR A LARGE COUNTRY PLACE

IV. Another example of a pressure tank supply system for a country estate is shown in Figs. 47 to 50. Fig. 47 shows the topography of a part of the grounds and gives the location of

the house and stable. The source of supply is a 6-inch driven
well about 200 feet in depth, yielding about 20 gallons per
minute. The water is pumped by means of a deep-well pump,
operated by a pump head driven by an electric motor. The
owner desired not only to use underground tanks in preference
to an elevated tank, but he also wanted the pump house to be
as inconspicuous as possible. With this in view the pump
house, shown in plan in Fig. 48, was built entirely underground,

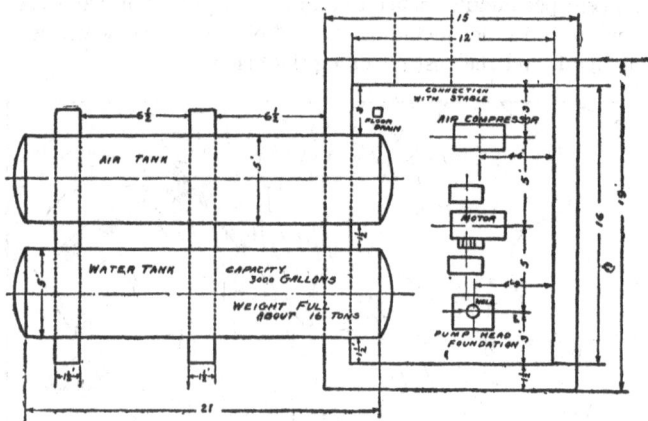

FIG. 48. — PLAN OF UNDERGROUND PUMP HOUSE, WITH PUMPING
MACHINERY AND TANKS.

nothing showing above the ground surface except a vent pipe
in an opening directly vertically over the well. This was
necessary in case the deep-well pump had to be taken out of the
well for repairs. Entrance to the pump house is by means of a
tunnel from the basement of the stable. The inside dimensions
of the pump house are 12 by 16 feet, and it contains a pump
head over the well, an air compressor and an electric motor,
with the necessary switch for turning on and controlling the
current. The two tanks, one for water, the other for compressed air, are located in the ground, and only their heads,
where the pipe connections are, project into the pump house.
Each tank is 5 feet in diameter and 21 feet long; the storage

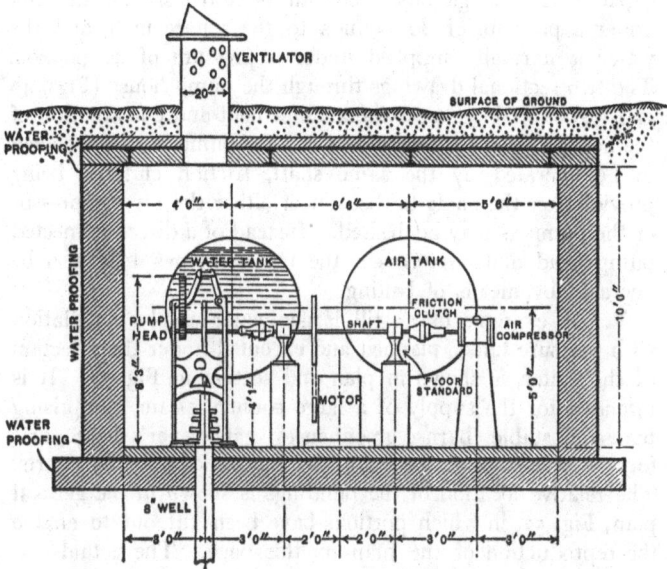

Fig. 49. — Cross Section through Pump House.

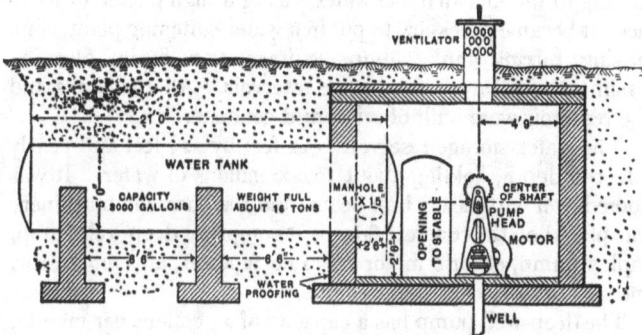

Fig. 50. — Longitudinal Section of Pump House.

capacity is 3000 gallons. The air is stored in the air tank under a pressure of 80 pounds to the square inch, and the water is normally supplied under a pressure of 40 pounds. The two sectional drawings through the pump house (Figs. 49 and 50) show in general the arrangement and construction of the house. The air compressor and the pump head are shown to be operated by the same shaft, friction clutches being provided on the shaft to disconnect either the air compressor or the pump as may be desired. Instead of a direct-connected pump head and compressor, the two machines may also be operated by means of belting.

V. An example of a still larger water-supply installation with pressure tanks, planned and executed under the direction of the writer, is shown in plan and section in Fig. 51. It is intended for the supply of a large country estate, comprising mansion, stable, barns, greenhouses, gatekeeper's lodge, and for the watering of the grounds and of lawn-tennis courts. The relative location of the buildings is shown in the general plan, Fig. 52, in which portions have been cut out to enable the reproduction of the plan on this page. The actual distances between the various buildings are greater than shown.

The supply of water is obtained from a driven and bored well about 258 feet deep, cased with 6-inch and 8-inch well-tube casing, the well yielding about 25 gallons per minute. Owing to the fact that the water was of a high degree of hardness, it became necessary to put in a water-softening plant, comprising mixing tank, sump, sedimentation basin, filtration basin and a clear-water storage reservoir. All the tanks and the reservoir were built of reinforced concrete.

The water-storage reservoir is 12 feet by 22½ feet and nearly 6 feet in depth, holding about 10,000 gallons of water. It was floored over with heavy iron beams and the pumping machinery was placed on the concrete floor. It comprises deep-well pump, triplex pump, electric motor of 20 horsepower, air compressor, and 20-horsepower gasoline engine.

The deep-well pump has a capacity of 25 gallons per minute, the triplex pump of 90 gallons per minute. Both pumps as well as the air compressor are direct-connected by means of shafting carried in pillow blocks, and provided with friction clutches to enable the throwing out of any part of the machinery. The gasoline engine drives a countershaft by means of belting, and the shaft is in turn connected with the pumps and the

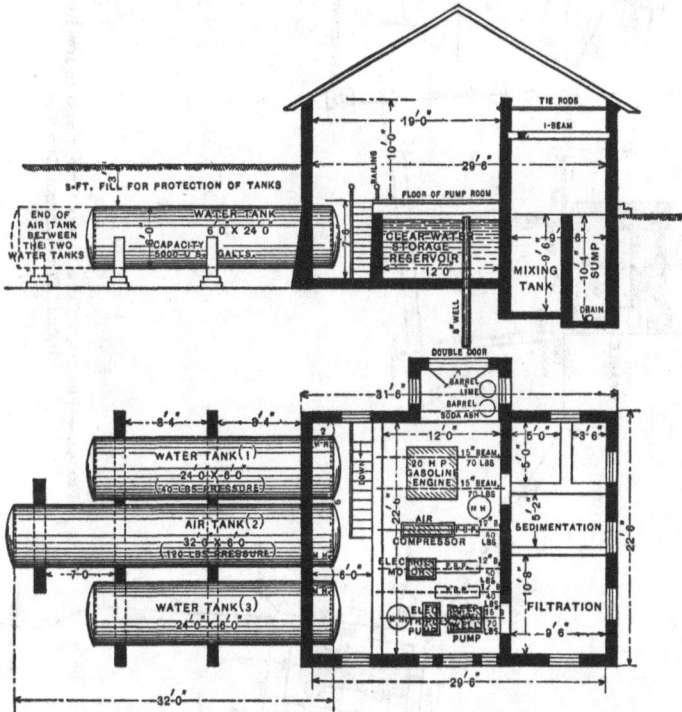

FIG. 51. — PLAN AND SECTION OF A WATER SOFTENING AND PRESSURE TANK PLANT FOR A LARGE COUNTRY ESTATE.

compressor. Should the electric power fail, the plant can be operated by the gasoline engine.

Water is stored under 40 pounds pressure in two water tanks, each 6 feet in diameter and 24 feet long, holding 5000 gallons each. Air is stored under 120 pounds pressure in one air tank, 6 feet diameter and 32 feet long. The tanks are placed in the

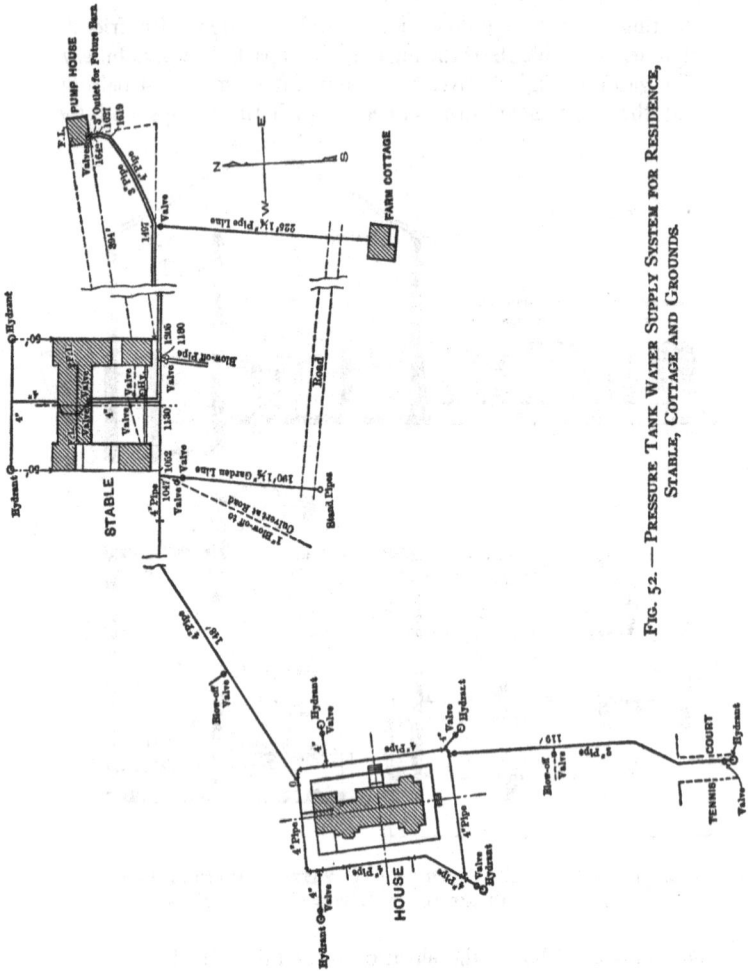

FIG. 52. — PRESSURE TANK WATER SUPPLY SYSTEM FOR RESIDENCE, STABLE, COTTAGE AND GROUNDS.

ground and are covered, but the heads of the three tanks extend into the basement of the pump room, which is reached by the stairs.

The well water is first pumped by the deep-well pump into the mixing chamber or tank, where soda ash and lime are added in the proportions required to render the water soft; it then passes through the sedimentation and filtration basins, and finally overflows into the clear-water reservoir.

The triplex pump takes its suction from this reservoir and forces the softened and clarified water into the pressure tanks. The air and water tanks are suitably connected and a pressure governor is provided, which is not shown in the illustration. It is similar to the one already described in connection with Fig. 46.

The pump house is about 24½ feet long and 31½ feet wide, and is entirely enclosed and provided with heating apparatus for the winter.

The water distribution to the buildings is as follows: A 4-inch water main, of cast iron "Universal" water pipe, is laid from the water tanks to the house, and supplies not only this, but also all the fire hydrants about the house and at the stable, seven in all, and also the fire lines of house and stable.

Another 3-inch line runs from the water tanks to the stable, and supplies all the plumbing fixtures of the stable and the plumbing in the farm superintendent's cottage. (See Fig. 52.)

VI. A water supply plant for a country estate at Port Chester, N. Y., which was put in under the author's directions, is shown in Figs. 53 to 58. The system adopted is the pressure tank system with separate air and water tanks.

Fig. 53 shows the location of the house, the garage, and of the pump house and water reservoir. The pumping apparatus was placed conveniently near to the reservoir, but the storage tanks for air and water were located on top of the hill, in the cellar of the garage building.

The dwelling house is located approximately 100 feet above the pump house. A duplicate plant is provided throughout, *i.e.*, two water tanks and two air tanks, each of 4000 gallons

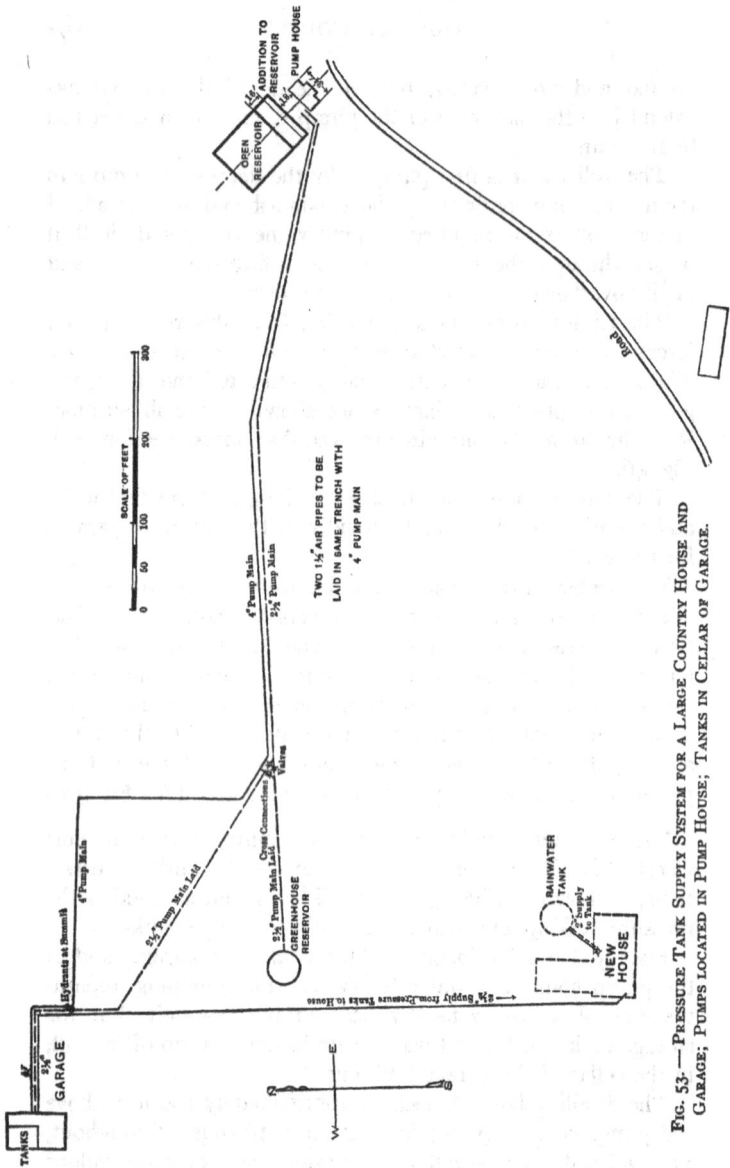

FIG. 53. — PRESSURE TANK SUPPLY SYSTEM FOR A LARGE COUNTRY HOUSE AND GARAGE; PUMPS LOCATED IN PUMP HOUSE; TANKS IN CELLAR OF GARAGE.

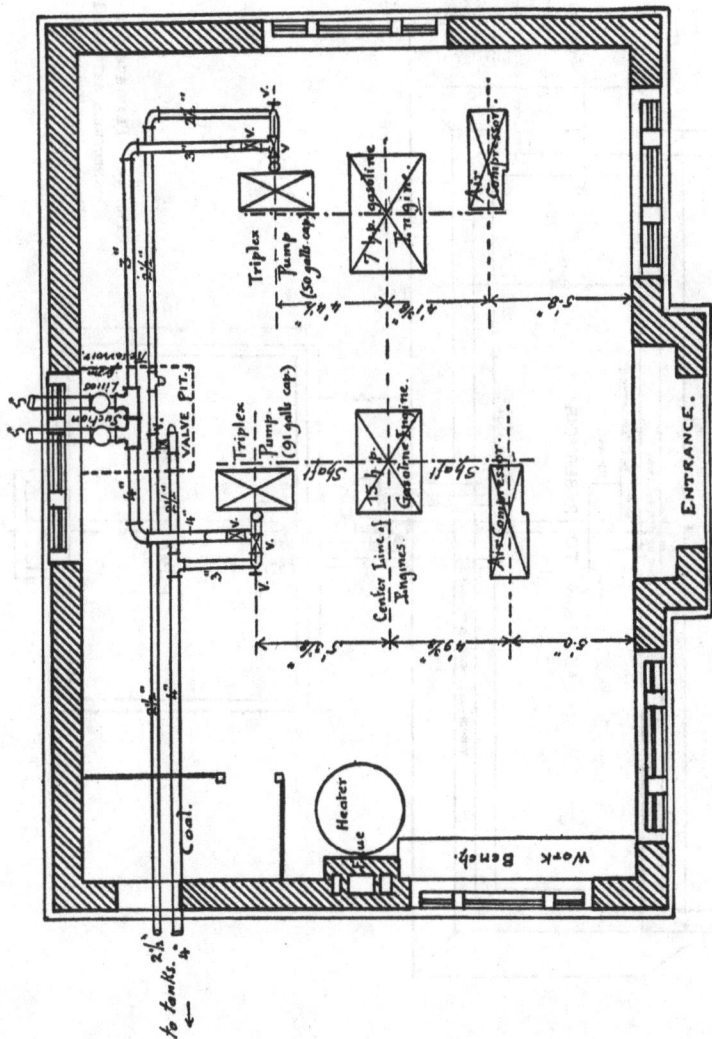

FIG. 54.—PLAN OF PUMP HOUSE, SHOWING DUPLICATE MACHINERY.

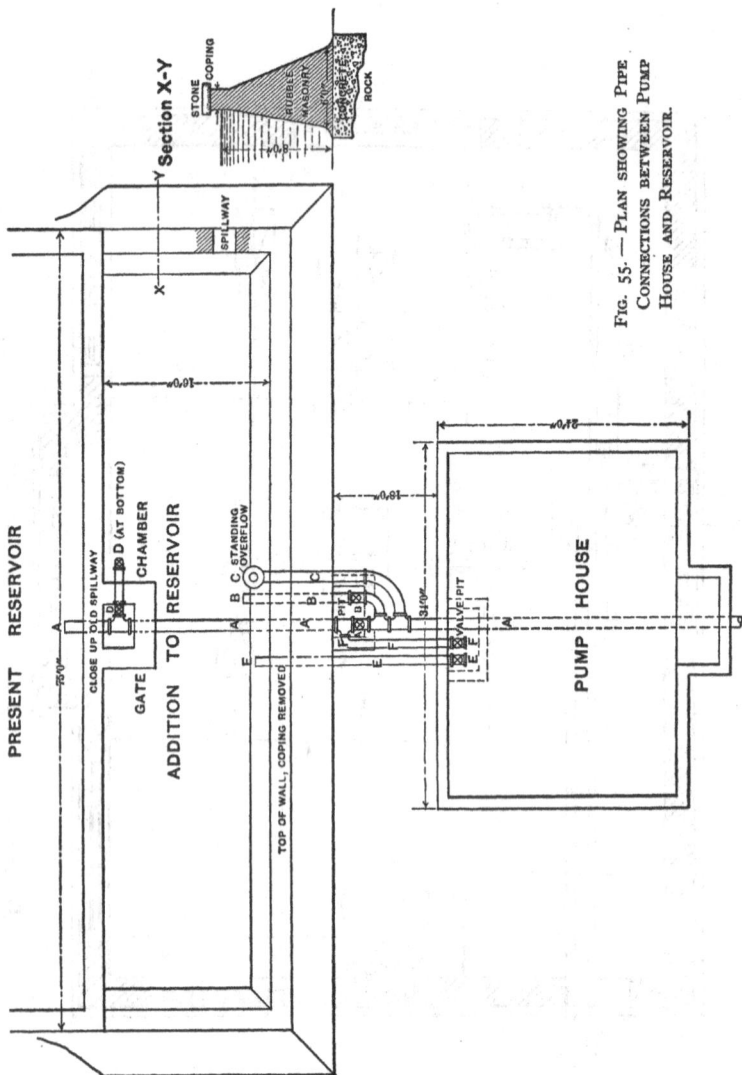

Fig. 55. — Plan showing Pipe Connections between Pump House and Reservoir.

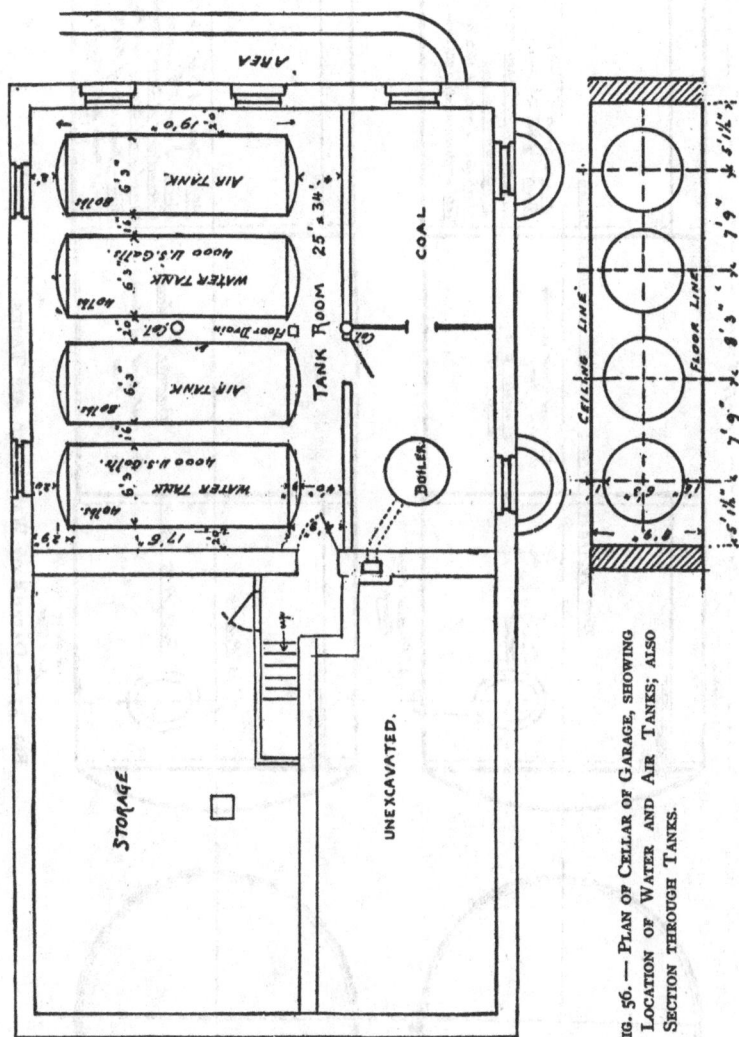

FIG. 56. — PLAN OF CELLAR OF GARAGE, SHOWING
LOCATION OF WATER AND AIR TANKS; ALSO
SECTION THROUGH TANKS.

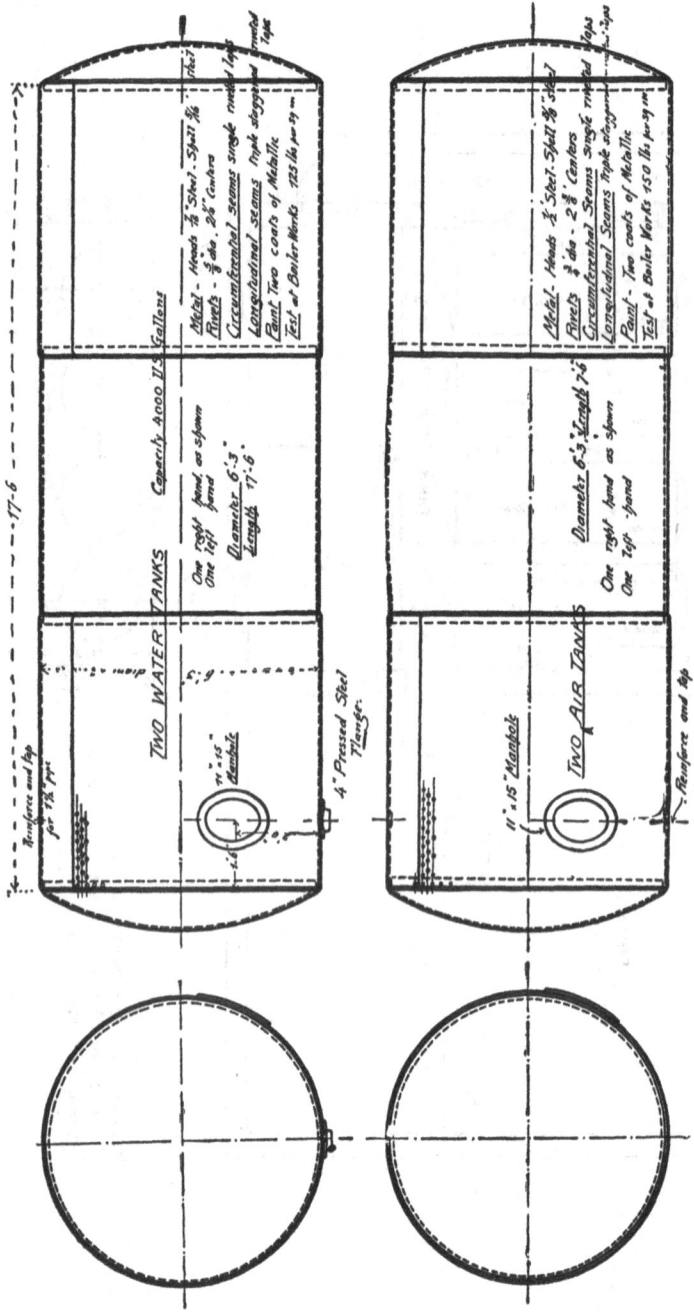

TWO WATER TANKS   Capacity 4000 U.S Gallons

Metal - Heads ⅜ Steel. Shell ⁵⁄₁₆ Steel.
Rivets - ⅞ dia., 2¾ Centers
Circumferential seams single riveted lap
Longitudinal seams Triple staggered riveted
Test at Boiler Works 125 Ths per sq in —

One right hand as shown
One left hand

Diameter 6'-3".
Length 17'-6

Reinforce and tap
for 1¼ pipe

4" Pressed Steel
Flange

11 x 15 "Manhole

TWO AIR TANKS   Diameter 6'-3. Length 7'-6

Metal - Heads ½ Steel. Shell ⅜ Steel
Rivets - ⅞ dia. - 2¾ Centers
Circumferential Seams single riveted laps
Longitudinal Seams Triple staggered
Paint - Two coats of Metallic
Test at Boiler Works 150 Ths per sq in

One right hand as shown
One left hand

11 x 15 "Manhole

Reinforce and tap
for 1¼ pipe

FIG. 57. — DETAILS OF WATER AND AIR TANKS.

capacity; the pumps and air compressors are also in duplicate. The machinery in the pump house comprises a triplex
pump of 91 gallons capacity per minute; another similar
pump of 50 gallons capacity; one Nash gasoline single cylinder engine of 7 horsepower and another larger double-cylinder
engine of 15 horsepower; also two air compressors, of 11 and
15 cubic feet free air capacity per minute.

A duplicate set of pumping mains is provided, viz., a 4-inch
pumping main for the larger pump, and a 2-inch main for the

FIG. 58. — VIEW IN PUMP HOUSE.

smaller engine and pump. Normally, the larger engine and
pump supply the tanks in the garage, and the smaller engine
and pump deliver water into a greenhouse tank, this being an
auxiliary system.

The position of the machinery in the pump house is shown
in Fig. 54. Fig. 55 shows the addition to the open reservoir
constructed under the author's directions and the pipe connections between it and the pump house. Fig. 56 shows the
location of the tanks in the cellar of the garage.

Fig. 57 shows details of the water and air tanks. Normally a
pressure of 80 pounds is carried in the air tanks and a pressure
of 40 pounds in the water tanks.

The two air pipes connecting the air compressors with the air and water tanks are laid in the same trench with the two water pipes. These air pipes are $1\frac{1}{2}$ inches inside diameter.

The two pumping mains are interconnected in the pump house and also at the garage. The piping for the reservoir and in the pump house pit is so arranged that water may be drawn either from the old reservoir or from the new addition. The work of construction was carried out by the Acme Water Storage and Construction Company. Fig. 58 is a view in the pump house taken by the author, and shows one of the gasoline engines with the triplex pump direct-connected to it.

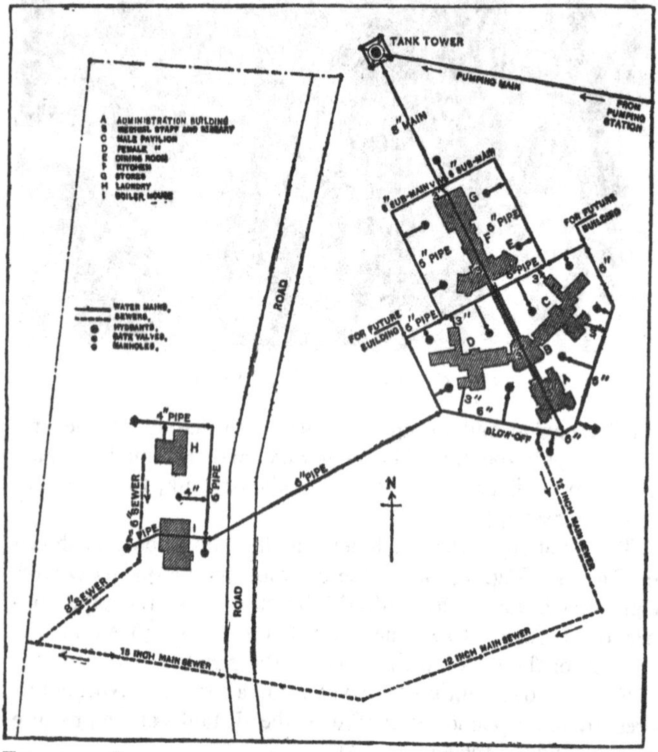

FIG. 59. — PLAN SHOWING ARRANGEMENT OF WATER DISTRIBUTION MAINS FOR A SMALL INSTITUTION.

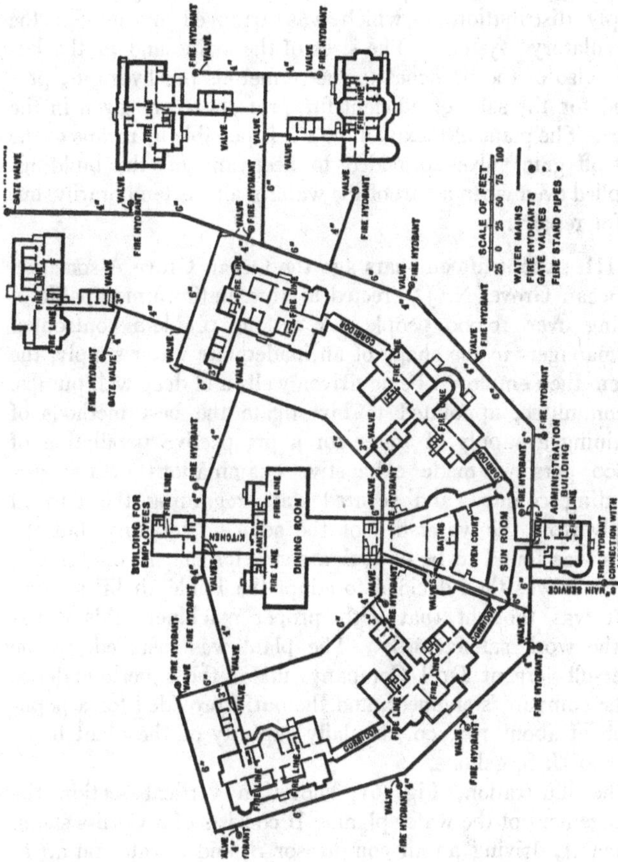

Fig. 60. — Plan showing the Water Distribution for a Large Institution.

VII. In order to illustrate the proper arrangement of the water-supply distribution system for larger institutions, I give in Figs. 59 and 60 two plans of groups of buildings, the water supply distribution of which was arranged by me on the "circulatory" system. The sizes of the mains and of the laterals, also of the branches to the numerous fire hydrants, provided for the sake of efficient fire protection, are given in the plans. The plans also explain how it is possible by means of the shut-off gate valves provided to keep any of the buildings supplied even when a part of the water mains is temporarily shut off for repairs.

VIII. About fifteen years ago the Ocean Grove Association of Ocean Grove, N. J., erected a large auditorium capable of seating over 10,000 people. A serious problem confronted the managers in the shape of an inadequate water supply, the system then employed being driven wells and deep well pumps. A committee, appointed to investigate the best methods of obtaining a supply of water for a prospective population of 50,000 persons, made exhaustive examinations of the surrounding country, and secured data regarding the natural supply from the watershed of the adjacent territory, but the water was found to be of such a character that it could not be used. It was then decided to adopt the Pohlé air lift system, as it was thought that with proper machinery this would do the work satisfactorily. The plant was installed by the Ingersoll-Sargent Drill Company, under the superintendence of the company's engineer, and the outfit provided for a population of about 100,000, the daily capacity of the plant being 250,000 U. S. gallons.

The illustration, Fig. 61, shows in vertical section the arrangement of the water plant. It consists of a Corliss steam engine A, driving an air compressor B, and a water pump F. The plant is made in duplicate, the engines being so proportioned that they may be run compound condensing. Steam is supplied by four return tubular boilers of 600 horsepower capacity.

By means of the Pohlé air lift pumps D, the water is taken

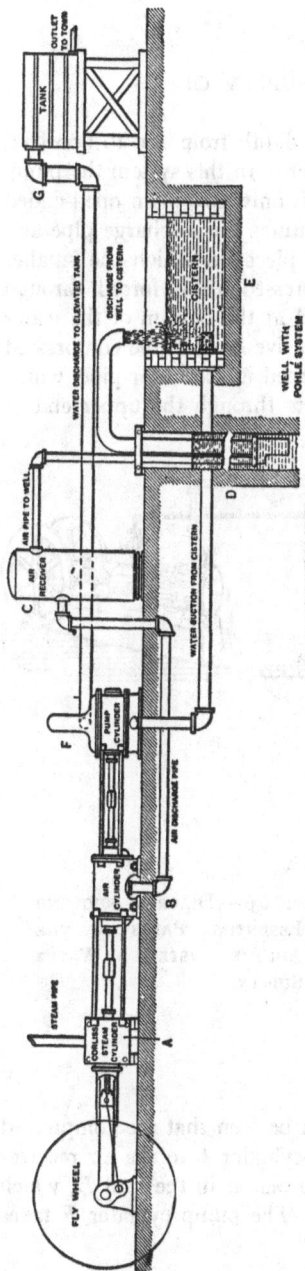

FIG. 61. — WATER SUPPLY BY MEANS OF THE AIR-LIFT SYSTEM.

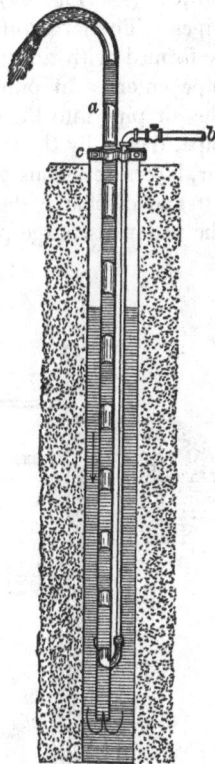

FIG. 62. — DETAIL SHOWING METHOD OF LIFTING WATER FROM WELL BY THE AIR-LIFT SYSTEM.

from 20 driven wells, ranging in depth from 400 to 600 feet, and from 4 to 6 inches in diameter. In this system the pump proper (see Fig. 62) consists of only two plain open-ended pipes. The larger of these constitutes the discharge pipe and is formed with an enlarged end piece, in which the smaller pipe enters. In pumping, compressed air is forced through the air pipe into the enlarged end at the bottom of the water pipe, thence by the inherent expansive force of the compressed air, layers or pistons of air are formed in the water pipe, which lift and discharge the water layers through the upper end of the water discharge pipe.

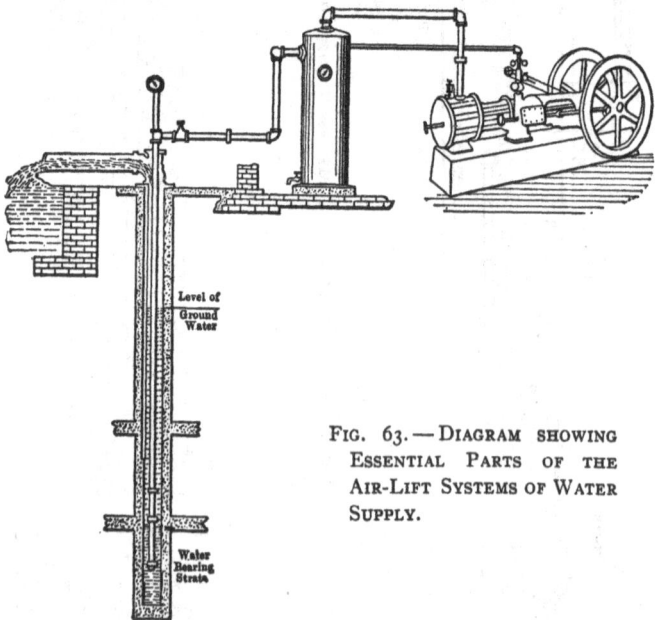

Level of
Ground
Water

Water
Bearing
Strain

Fig. 63. — Diagram showing
Essential Parts of the
Air-Lift Systems of Water
Supply.

Upon examining Fig. 61 it will be seen that the compressed air passes from the compressing cylinder B to the air receiver C and thence to the Pohlé pump placed in the well D, which lifts the water to the cistern E. The pump cylinder F takes its supply from the bottom of the cistern and delivers it into the tank *G*, from which it enters the distributing mains. The plant thus consists of a combination of two systems — by means of compressed air the water is first raised to a cistern at the surface, and from thence it is raised by a steam pump to the distributing tank. The plant has been found economical and reliable in every respect, and no trouble whatever has been experienced in obtaining all the water required. — (*From the Metalworker*).

Fig. 63 is a diagrammatic illustration of the method of operation of the air-lift system. This cut was intended to be inserted on page 147, in the general description of the air-lift method of pumping water, but was omitted by an error in printing.

IX. The following description of a water works installation for the Narragansett Bay Coal Depot, of the U. S. Navy

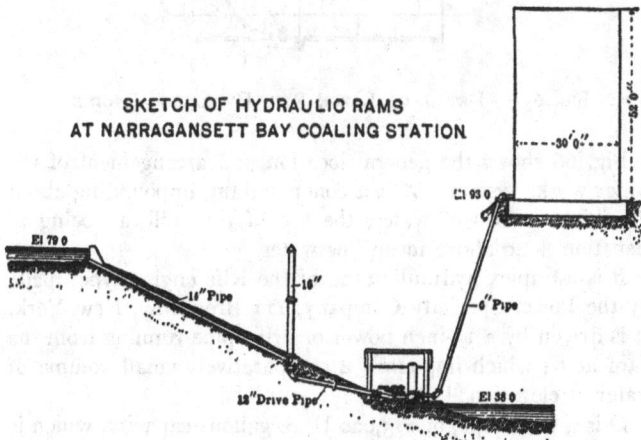

FIG. 64. — WATER SUPPLY BY MEANS OF A RIFE HYDRAULIC ENGINE.

Department, in Narragansett Bay, R. I., is condensed from a paper by Augustus Smith, Mem. Am. Soc. C. E., from Transactions, Vol. LVII, Paper No. 1032 of 1906.

The Bureau of Equipment installed the plant, which is shown in Figs. 64 and 65.

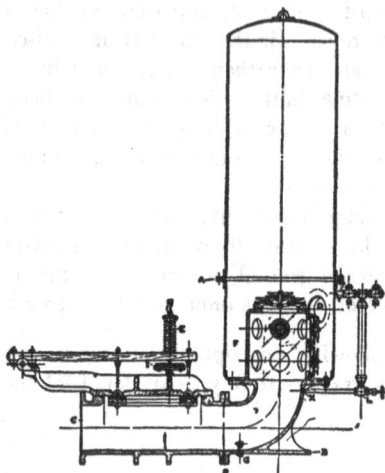

FIG. 65. — DETAIL OF LARGE RIFE HYDRAULIC ENGINE.

Fig. 66 shows the general location and arrangement of the water works system. *A* is a concrete dam, impounding about 3 million gallons of water, the top of the spillway being at elevation + 40 above mean low water.

B is a duplex hydraulic ram, of the Rife engine type, made by the Power Specialty Company, 111 Broadway, New York. It is driven by a 12-inch power or drive pipe running from the dam at *C*, which impounds a comparatively small volume of water at elevation + 81.

D is a standpipe, of 275,000 U. S. gallons capacity, which is filled by an 8-inch delivery pipe from the ram. The remainder of the water wastes into the lower reservoir. The bottom of the standpipe is at elevation + 93 above mean low water. The standpipe is 30 feet in diameter and 52 feet high. *E* is a cast-iron pipe supplying the fire hydrants and water supply system at the station.

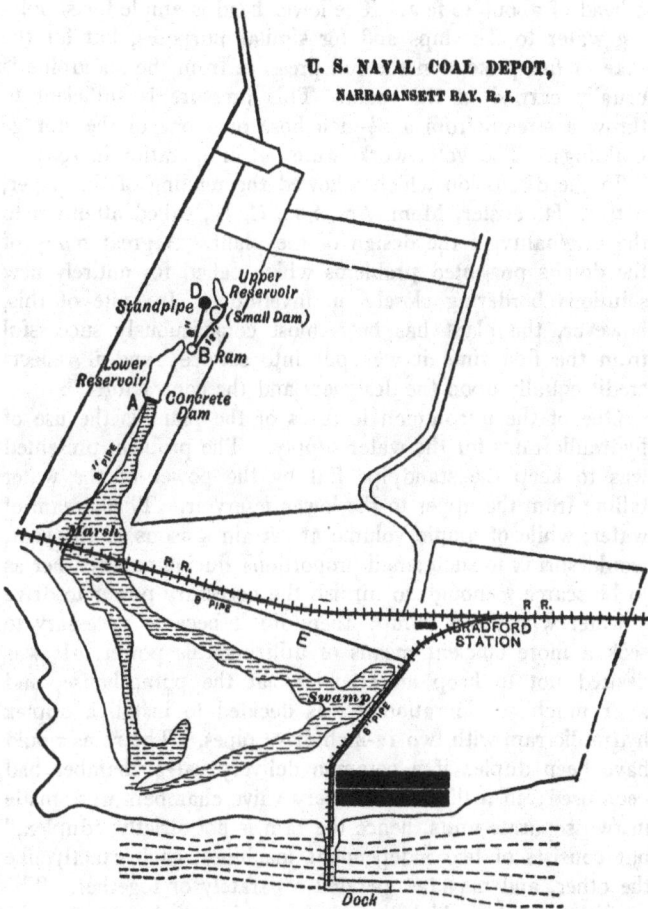

FIG. 66. — PLAN SHOWING WATER SUPPLY BY RAM FOR UNITED STATES NAVY DEPARTMENT'S COAL DEPOT AT NARRAGANSETT BAY

By suitable valves water can be drawn either from the standpipe under the full head, or from the main reservoir under a head of about 40 feet. The lower head is ample for supplying water to the ships and for similar purposes, but for the sake of fire protection the full pressure from the standpipe is usually carried in the main. This pressure is sufficient to throw a stream from a 2½-inch hose over one of the storage buildings. The water works were put in operation in 1903.

In the discussion which followed the reading of the paper, Mr. E. H. Foster, Mem. Am. Soc. C. E., called attention to the originality of the design of the plant. A great many of the details presented problems which called for entirely new solutions bordering closely on inventions. In spite of this, however, the plant has been most conspicuously successful from the first time it was put into service, and it reflects credit equally upon the designers and the contractors.

One of the uncommon features of the plant is the use of hydraulic rams for the water supply. The problem presented was to keep the standpipe full by the power of the water falling from the upper to the lower reservoir. This stream of water, while of ample volume at certain seasons of the year, would shrink to such small proportions during dry weather as to be scarcely enough to furnish the necessary power to drive a water wheel and pump; therefore it became necessary to seek a more efficient means of utilizing this power. It was desired not to keep an attendant at the pump house, and after much consideration it was decided to install a duplex hydraulic ram with two 12-inch drive pipes. The rams would have been duplex if a common delivery valve chamber had been used; in reality, the delivery valve chambers were made in two separate units, hence the ram is not strictly "duplex," but consists of two independent machines, each exactly like the other, and may be operated separately or together. The conditions under which the rams were installed were that the efficiency should be 70 per cent; also that they should be capable of filling the standpipe in 36 hours when the power water was supplied at the rate of 1 cubic foot per second, or in 12 hours with 3 cubic feet per second.

The results obtained in the test are interesting, as they exceeded all the guaranties considerably; in fact, they were quite remarkable; they laid bare the hitherto undeveloped possibilities of this kind of hydraulic machine for pumping water, where water power exists, without attendance and with high efficiency. These rams were designed specially for this work, and while perhaps not the largest that have ever been built, were probably the first large rams to be used with such high working head. It would appear, from the results which these have given, that rams of much larger size could be built with safety.

Extract from Official Test of Rams.

	No. 1	No. 2
Total water delivered to engine, in gallons per minute (Q)	582	578
Water delivered to standpipe, in gallons per minute (q)	232	228
Power head, in feet (H)	36.75	37.25
Pumping head in feet (h)	47.25	46.85
Strokes per minute	130	130
Efficiency per cent (D'Aubuisson formula)	91.11	88.95
Efficiency per cent (Rankine formula)	85.22	81.76

In connection with the test of the rams, an interesting discussion as to the proper formula to be used in determining the efficiency developed. Two well-known authorities, Rankine and D'Aubuisson, each give formulas which differ materially, thus:—

Rankine formula:

$$E = \frac{qh}{(Q - q) H};$$

D'Aubuisson formula:

$$E = \frac{q (H - h)}{QH}.$$

Under these conditions of fall and pumping head, the second formula gives an efficiency nearly 7 per cent greater than that of the Rankine formula.

After some correspondence the Department finally ruled that the D'Aubuisson formula was the more logical, and therefore should be accepted as correct, on the grounds that only the rams were actually under test, and that they should be charged with the energy received at the point of reception, and credited with the energy delivered at the corresponding point of delivery. In whatever way this matter may be regarded, the efficiency obtained is certainly very high, and cannot be approached by any other form of hydraulic motor known. From this it would seem as if a very large field of usefulness awaits the further development of hydraulic rams.

NOTE. — Bulletin No. 205 of the University of Wisconsin, of March, 1908, brings a dissertation on "An Investigation of the Hydraulic Ram," by Leroy Francis Harza, which is very interesting, and in which the author adopts the Rankine formula as the one from which to figure the work performed by the ram.

X. Fig. 67 shows an interesting example of the modern use of concrete in water supply structures. The water tank shown was designed by Mr. F. J. Sterner, architect, and built by Mr. DeLancey A. Cameron, to whom I am indebted for the following description of the structure, taken from an article contributed by Mr. Cameron to *Cement Age* of January, 1908.

The tank gives a water storage of approximately 15,000 U. S. gallons, and is designed to supply a residence and grounds at Katonah, N. Y. The tank is 22 feet outside diameter, with a circular wall, 10 inches thick and 6 feet deep inside, sloping several inches toward the center. It stands on eight concrete columns placed at the circumference, with concrete girders extending from four of these columns to a center column. The floor is 12 inches thick at circumference and about 8 inches at center.

The concrete was formed in about the proportion of 1 cement, 2 sand and 3 broken stone, and used very wet for both the floor and the walls. The wall is reinforced with Clinton welded reinforcement ⅜-inch wire, 2 by 12 mesh, extending around the circumference in one length and fastened together at the ends. It is placed about 2 inches from the outside. The girders, 18 inches by 12 inches, have ½-inch square rods at the bottom, and Clinton reinforcement extending from six inches in the floor, along the outside of the girder under the rods and along the other side, and 6 inches into the floor again. The floor is reinforced with same kind of mate-

FIG. 67. — VIEW OF A REINFORCED CONCRETE WATER TANK.

rial running across the girders and into the walls, where the ends are turned up. The supply and waste pipes extend through a hole in center column. The tank was stippled with cement mortar on the outside, and then one coat of a cold water paint was applied. On the inside it had a coat of cement mortar, 1 cement to 2 sand, plastered on the concrete of wall and floor, and then two washes with very thin neat cement applied with brush, this being done several weeks after concrete was formed.

The tank without a roof had 4 feet of water in it last winter, when cold weather came on, and ice formed 10 inches on the top and over 2 inches thick on the sides down to the bottom. The tank showed no cracks and needed no repairs after the ice melted in the spring. It is absolutely water-tight and shows no dampness on the outside of walls or floor.

The builder states that he inspected the tank recently and found it in perfect condition, except that the white paint on the exterior had more or less fallen off. He recommends to use white cement instead of white paint. It was found necessary during the winter to box in the central column of the tank through which the intake and outlet pipes run, to keep them from freezing, because the tank is located in a very exposed position at the top of a hill.

Mr. Cameron makes a specialty of buildings in concrete and is at present constructing several other reinforced concrete reservoirs.

XI. The following work of water supply for a country estate, designed by Mr. Albert L. Webster, C.E., consulting engineer, is of interest. The estate, located at Bernardsville, N. J., comprises about 300 acres of land, hilly and well wooded, which was laid out as a private park, under direction of a landscape architect, and which also contains farm and grazing land, a kitchen garden and other accessories of a country villa. At the summit of the principal hill (see Fig. 68) a large mansion was erected, and at convenient locations a stable, a farm house, a dairy, barn and employees' houses. The estate was further improved by grading, drainage, the building of various roads and walks, the provision for irrigation, and the improvement of a natural lake.

The water supply is obtained from several springs, and its collection, storage and distribution, together with the disposal of the sewage and of rain water, involved considerable engineering work, executed under the direction of Mr. Webster.

The spring from which the entire water supply is obtained is located at *A*, Fig. 68. This spring has a minimum daily flow of about 8000 gallons. Its flow is stored in a twin res-

ervoir R, of 70,000 U. S. gallons capacity. The details of
this reservoir are shown in Fig. 69. The walls are built of
rubble masonry, laid in Rosendale cement mortar. The
bottom is lined with a bed of concrete, 8 inches thick,
made continuous with a footing under the walls. The entire
interior surface is plastered 1 inch thick with Portland cement

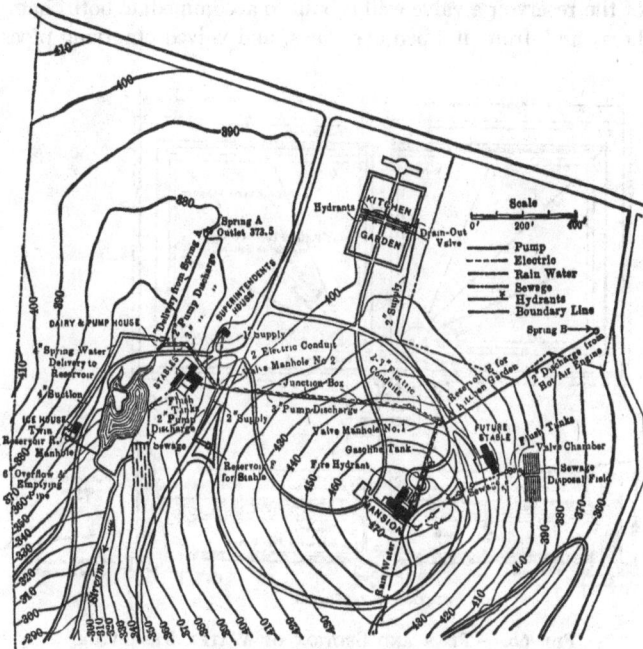

FIG. 68. — PLAN OF COUNTRY ESTATE SHOWING WATER SUPPLY
AND SEWAGE DISPOSAL SYSTEMS.

mortar, mixed in the proportion of 1 sand to 1 cement. The
side of the reservoir adjoining the ice house is coated with
asphalt and the entire reservoir is surrounded with footing
course tile drains.

The reservoir is divided into two equal chambers by a
central partition wall, and it is filled from the spring through a

4-inch pipe. This pipe has a valved branch discharging into the top of each chamber at the corners opposite the overflow pipes. The supply from the reservoir is pumped through a vertical suction pipe in each chamber provided with gate valves, foot valves, strainers and emptying valves. The suction pipes connect with a common pump main. On one side of the reservoir a valve well is built to accommodate both chambers, and from it open overflows, and valved emptying pipes

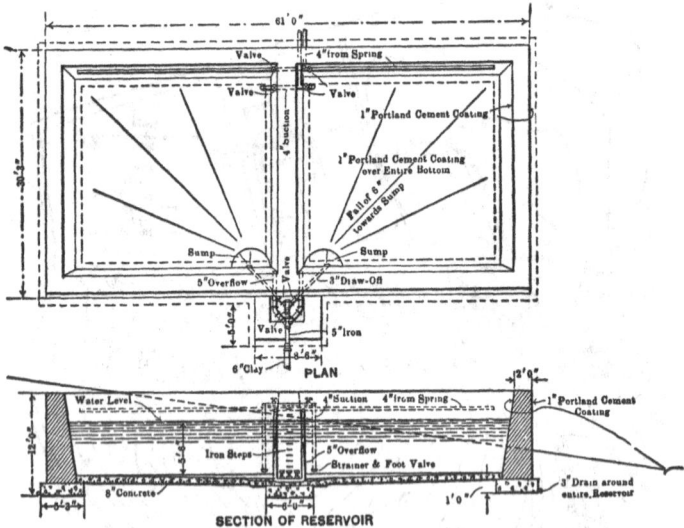

Fig. 69. — Plan and Section of Water Reservoir.

are brought from the respective chambers and terminate in connections to an iron pipe which is extended beyond the line of the reservoir with 6-inch galvanized iron pipe and there connected with a 6-inch clay pipe discharging into the adjacent stream, which formerly received the entire flow from the spring. The emptying pipes are valved so that they can serve also as an equalizing connection between the two divisions of the reservoir.

The reservoir is protected by a simple wooden roof, and the masonry walls have an embankment sloped down on the outside, and turfed over. Adjacent to it is an ice house about 50 feet square for the storage of ice cut on the lake.

Another spring *B*, Fig. 68, is provided with a hot air pumping engine, and can furnish 20,000 gallons more water a day if required, and a third, having a capacity of 17,000 gallons daily, can be made available if necessary.

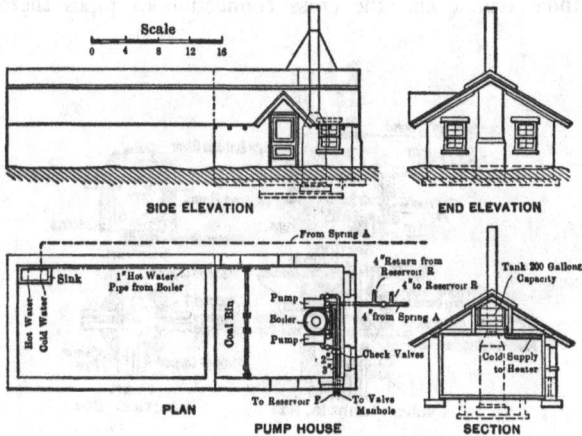

Fig. 70. — Plan, Section and Elevations of Pumping Station.

The main supply is pumped from reservoir *R* by duplicate steam pumps located in a pump house, which normally deliver to the tanks at the mansion and to the elevated reservoirs *E* and *F*, and can also throw a fire stream over the house. The reservoir *E* has a capacity of 17,000 gallons, and is intended to provide for garden irrigation. Reservoir *F* has a capacity of 3000 gallons, and supplies the stable and farm buildings.

A hot-air pumping engine is installed at the spring *B*, and its discharge pipe is cross connected so as to deliver directly to reservoir *E*, or to the main from the steam pump at the dairy house, and thus to all the reservoirs and tanks.

The pump pipes are all carried in trenches 4½ feet below the surface of the ground, which also contain electric conduits for light and telephone wires, and which are accessible through brick manholes at the valve boxes, junction boxes, etc.

The arrangement of the steam pump and the connections of the pipe at the main pumping station in one end of the dairy house are shown in Fig. 70. The construction of valved manhole No. 1 and the cross connection of pipes there is

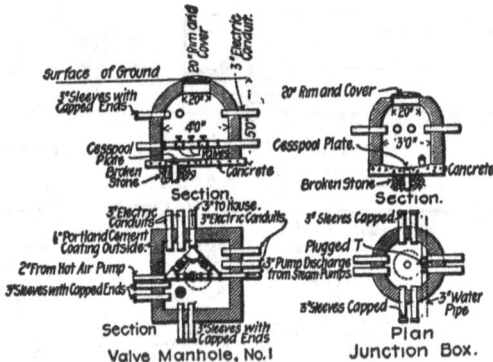

FIG. 71. — DETAILS OF MANHOLES AND JUNCTION BOXES.

shown in Fig. 71. Here, as in all manholes, several capped sleeves are built in the wall in order to provide for future electrical connections if desired, without cutting the brick-work. Plugged Tees are also provided on water pipes in each manhole intended for future extensions if required. One of the junction boxes on the main line is shown in Fig. 71; it is also provided with capped sleeves and a plug on the water pipe, which permit future extensions. [From *Eng. Record.*]

The sewage disposal plant of this estate is described in the examples given in the third part of the book.

III.
SEWAGE DISPOSAL

THE SEWAGE DISPOSAL OF COUNTRY HOUSES

THE SEWAGE QUESTION FOR ISOLATED COUNTRY HOUSES AND VILLAGE DWELLINGS *

Evils of Cesspools and Privy Vaults. — It is a pernicious and barbarous custom to store the liquid household wastes and human excreta for any long period of time near habitations. If thus imperfectly disposed of, they soon undergo a most dangerous putrefaction, become the cause of a pollution of the soil, the air and the water, and may favor the breeding and spreading of infectious diseases. Primitive methods, however, are gradually, though slowly, giving way to more judicious methods of sewage disposal.

Great centers of population have long ago decided to abolish the dangerous cesspools and the filth-reeking privies; even many of the smaller cities and towns are waking up to the necessity for immediate action. Less perceptible, though undoubtedly not wholly lacking, is the interest taken by the smaller village communities, or by isolated country estates and farmsteads, in this question of a proper disposal of the liquid household refuse.

* Sewerage and sewage disposal for country houses are discussed at length in the author's book "The Disposal of Household Wastes," also in his book "Sanitary Engineering of Buildings." The three articles reprinted here appeared originally in various technical publications, and give a broad general review of the subject. The last one includes a more detailed discussion of bacterial methods of sewage disposal, and contains many plans and illustrations taken from the practice of the author, as well as some from that of other engineers.

W. P. G.

The remedies to be applied vary necessarily for various localities and for different conditions. In the case of cities, of towns and of densely populated villages, it is obvious that only the *united action* of the residents can effect any reform. In other words, the community as a whole must devise proper measures for sewage removal and disposal, by employing expert engineers and specialists to design and construct the system decided upon.

Sewage Disposal for Detached Houses. — In rural districts, where houses are isolated and generally at long distances apart, the case is quite different. Here each owner is, generally, compelled to meet alone the difficulties which confront him and to make the required provisions for his own sewerage and sewage disposal. Fortunately, these are not often insurmountable, as the following considerations will show.

Isolated country dwellings and farmhouses usually derive their water supply from wells, springs or cisterns. It is, consequently, of the utmost importance to keep the water pure and to prevent any possible contamination, either directly or by soil pollution. It is clear, at the outset, that the first imperative duty is to do away with any privy or leaching cesspool that may exist on the premises.

Earth Closets. — A simple and cleanly, and in all respects entirely satisfactory substitute for the privy is the earth closet, of which inexpensive, as well as more complicated types exist. The action of dry earth, in not only deodorizing, but also rendering harmless, excreta of men and animals, has long been well known. More recently the observation was made that if the urine be kept separate from the solids, a much smaller quantity of earth will be required to cover and absorb the latter, also that the closet will be more easily kept free from smell, and that it may therefore be located close to the dwelling without becoming a nuisance, if it only be properly used and well taken care of. The closet seat, therefore, should be arranged in such a manner that a separation of solids and liquids is at once effected. The dry earth manure ought to be removed at frequent intervals, and used in the garden attached to the country house. But what to do with the urine after separation is the next question?

Slop Water Disposal. — Chemical researches have long established the fact that the most valuable fertilizing elements are contained in the liquid manure, and for this reason alone the aim should be to utilize it. Inasmuch as every country house, no matter how small, must necessarily dispose of another kind of liquid refuse, comprising the slop water from bedrooms, the soapy wastes from the wash tubs, and the greasy waste water from the kitchen sink, the thought readily occurs to combine the urine with the slop water. The fact is well known to every gardener that soap suds make an exceedingly valuable liquid manure for pot plants, vegetables and fruits of all kinds.

The remedy for the usual slop-water nuisance in villages or farmhouses is, therefore, obvious. Instead of throwing the slops on the ground near the kitchen door, where they soon create a bad nuisance, and instead of allowing the filthy liquid to soak away from a leaching cesspool into the subsoil, all slop water should be collected in a small underground tank, built thoroughly tight, and the liquid should be distributed either by gravity or by means of a pump, at frequent intervals in the kitchen garden, or used to irrigate the roots of shrubbery, or, if at some distance from the house, a small surface irrigation system may be arranged.

Inasmuch as a slop-water tank will be needed for even the smallest cottage, to temporarily store the liquid sewage, it is a simple matter to run the urine from the earth closet to it by means of a small drain pipe. To prevent any obstruction it is advisable to insert into the funnel under the closet seat a sieve or strainer.

Subsurface Irrigation. — A more elaborate, yet simple method of disposing of slop water is that by means of "subsurface irrigation," and this has been recently introduced in many isolated country houses. In this system the liquid sewage, after being collected in a tight vessel or sewage tank, is distributed at intervals under the surface of the soil by a network of small, porous, open-jointed drain tiles, and the sewage is purified by the action of vegetation, the roots of grass or shrubbery taking up nourishment from the liquid, while the latter filters away through the soil, in which bacterial purification takes place. The method is equally applicable to those country houses which, having an abundant supply of water, are furnished with water-closets and bathrooms, but the details of the arrangement require in this case some modification. While not always *perfect* in its action, it is one of the best methods of removing and disposing of the sewage in the case of isolated country houses.

In this connection the following quotation from an early volume of the *Sanitary Engineer* is interesting:

"An open cesspool for waste water from bath and basins only will contaminate the soil about it quite as surely as if receiving the drainage of water-closets, though perhaps not quite as rapidly. It is better to distribute the refuse on or near the surface. A collection of waste water from bath tubs and washbowls will become as foul as any other refuse from the house, and differs only from other refuse in being diluted by a greater bulk of water.

" The disposal of house sewage in suburban residences is a very troublesome question, especially where a public water supply is afforded without sewers — a state of things which is not compatible with public safety. Distribution *on* the surface is the simplest method, but this requires at least an acre about every house, in order to make it free from offense.

" It can be distributed in porous tiles about 10 inches *under* the surface on smaller house lots if the conditions are favorable — viz: a well-drained or porous soil.

" Whether on the surface or beneath it, there must be slope enough to allow the sewage to flow by gravity from the bottom of the cesspool to the place of distribution, otherwise pumping is necessary, which is, of course, onerous.

" When distributing in pipes beneath the surface, it is usual to lay 2-inch unglazed tiles, with joints at least one-fourth of an inch open, in trenches 10 or 12 inches deep, graded with a uniform fall of not over 1 inch in 10 feet at most, and sometimes not over half this amount. Sufficient ramification and length of pipe must be provided to give a capacity of at least one-half that of the cesspool or tank to be discharged; the other half is generally soaked away during the flow. It is important to secure an intermittent flow in the tiles. If a constant driblet flows into them they will always be chocked in a short time. When on a larger scale, as for hotels and public institutions, Field's flush tank is used with success to secure this periodic or intermittent flow. If the water supply is limited to what is pumped by hand in the house, the discharge can be made by means of a stop-gate laid in the outlet pipe leading from the bottom of the cesspool, and operated by hand when required. The tiles adapted for this purpose are readily obtainable."

Sewage Tanks and Ordinary Cesspools Compared. — It may be claimed that a slop-water or sewage flush tank, no matter how small and how well built, remains in some sense and to some degree a cesspool, because the liquid sewage is retained in the tank until disposed of by surface or subsurface irrigation. But in practice it is found that, if such a tank is frequently emptied, properly cleaned, occasionally disinfected, and if the distribution is effected with regularity, the system will, with some care and a little attention, work satisfactorily and without becoming a nuisance. It requires a garden, lawn or meadow of small area near the dwelling.

Examples of Simple Subsurface Irrigation Systems. — Two examples of cheap sewage disposal arrangements for smaller houses are illustrated in Figs. 72 and 73, and

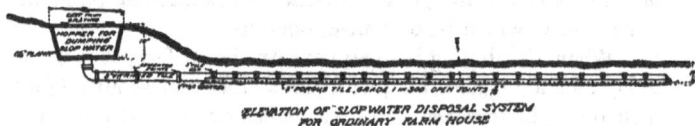

ELEVATION OF "SLOP WATER DISPOSAL SYSTEM
FOR ORDINARY FARM HOUSE

FIG. 72. — A SIMPLE SEWAGE DISPOSAL ARRANGEMENT.

the methods indicated therein are so simple as not to require any further explanation.

Sewerage for Village Houses. — In closely built up villages, with no space about the house available for irrigation, a sewage disposal system on the premises is, of course, inapplicable. The only remedy in such a case is to build a main pipe sewer — a 6-inch pipe answers for a whole village of 1000 or more inhabitants — with smaller branch pipes to each dwelling for the removal of the slop water, without the admission of excreta, and to carry the sewage well beyond the limits of the village.

Whether its immediate discharge into some stream be permissible, or whether a previous purification by sedimentation, screening, chemical precipitation, land filtra-

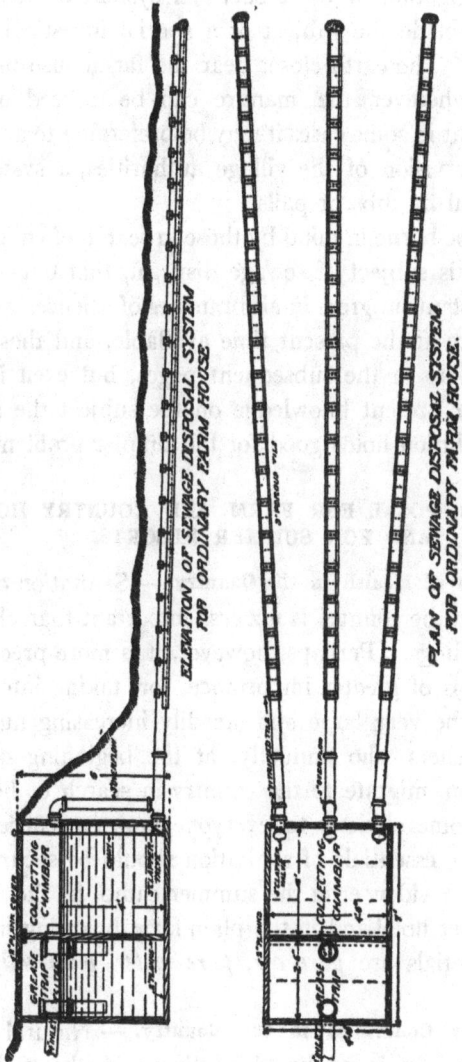

ELEVATION OF SEWAGE DISPOSAL SYSTEM
FOR ORDINARY FARM HOUSE.

PLAN OF SEWAGE DISPOSAL SYSTEM
FOR ORDINARY FARM HOUSE.

FIG. 73.—A SIMPLE SYSTEM OF SEWAGE DISPOSAL.

tion, by irrigation or by a bacterial system be required, should be made the subject of a special investigation in each case. The earth closet near the farmhouse may be retained, wherever the manure can be utilized on the grounds, but in some cases it may be preferable to arrange, under supervision of the village authorities, a system of dry removal by tubs or pails.

It must be borne in mind by those in search of enlightenment on this subject of sewage disposal, that ours is the age of constant progress in all branches of science. Other methods are at the present time available, and these will be referred to in the subsequent pages, but even in the light of our present knowledge on the subject the above outlined solution holds good for the simpler problems.

SEWAGE DISPOSAL FOR FARM AND COUNTRY HOUSES AND FOR SUMMER RESORTS.

Essentials of Health in the Country. — Sanitation on the farm and in the country is not less important than that of urban dwellings. Perhaps, however, it is more precise to say that it is of greater importance, for, taking into consideration the very large and steadily increasing number of city dwellers who annually, at the beginning of the warm season, migrate to the country in search of health, it does become obvious to everyone how very necessary it is that the essentials of sanitation should be secured in the country residence, in the summer camp, in the luxurious summer hotel and in the plain farm boarding house. These essentials are *pure air, pure water, pure soil,* and *pure food.*

Unsanitary Conditions in the Country. — Natural conditions certainly favor healthful living in the country.

But on the other hand many artificially created conditions tend to affect unfavorably the health and welfare of the farmer or the dweller in the country and his family. Indeed some statistical figures seem to prove that there is more illness and suffering, more preventable disease, in the country than in the city. In many localities the soil is unduly saturated with water, the ground is imperfectly drained, the surroundings are undesirable and suggest the existence of malaria, the house is badly located, the water supply is contaminated, the sewerage arrangements are of the crudest and worst kind, and there is generally an absence of cleanliness and neatness, and on the contrary much dirt in one form or another. In many cases a non-compliance with obvious health laws forms the cause for the prevailing illnesses, and it is often not difficult to trace outbreaks of typhoid fever or other diseases to a general lack of sanitation. On many farms, flies, mosquitoes, and rats abound, and these, it is well known, may act as the carriers of infection.

Sanitary Conveniences for Farmers. — A certain improvement is, however, noticeable at the present time in the sanitary conveniences of farmers' homes. This improvement is the natural outgrowth of the increased requirements for comfort and refinement in the home. The electric light, the long-distance telephone, the suburban trolley lines, and other nineteenth century inventions are having an ever-increasing tendency to revolutionize modern life on the farm. Until recently, sanitation has been lagging behind, but an awakening in this direction is noticeable. Improved water supply and proper sewage disposal are destined to become important factors in the farming enterprises of the near future.

The average farmer too often entertains the wrong idea that such improvements are extremely expensive. Such, however, is far from being the case, and if kept within the bounds of utility, avoiding anything at all luxurious, the cost of the installation of plain sanitary fixtures is considerably less than might be anticipated.

Surely no good reason exists why the benefits of a pure water supply, suitably installed under a good pressure, of convenient and time- and labor-saving plumbing fixtures, of a plain, clean and sanitary indoor water-closet, and of good and efficient drainage, should not be extended to the cottages of the farmers as well as to the country mansions of the rich.

A residence on a farm, or on a country estate where health laws are observed and where the ounce of prevention is applied in matters of drainage, water supply and waste disposal, must necessarily be conducive to health and longevity.

Water Supply, Sanitary Plumbing, and Waste Disposal. — If the farmhouse is to have plumbing fixtures, three things must be provided; first, an adequate water supply; second, safe and sanitary plumbing; and third, proper means of disposing of the wastes from the household.

The provision of a good and sufficient supply of water has been discussed in the second part of the book, whereas the installment of plumbing appliances was briefly dwelt upon in Part I. The question of the safe disposal of the sewage is surely the most frequent and most serious problem which confronts the farmers, the dwellers in country houses, the superintendents of isolated institutions, and the managers of summer resorts. Notwith-

standing this fact, the problem is one to which comparatively few people pay sufficient attention.

Sewage Disposal. — A haphazard mode of disposing of
the sewage is no longer permissible, for sewage not
properly taken care of soon becomes offensive, causes a
nuisance to sight and smell, and may become a source of
actual danger to health if it contains the germs of disease
and if it contaminates a well, spring or a stream used for
drinking water.

Many scientific experiments on sewage purification on
a small as well as on a large scale have recently been
made, but the results of the investigations have not
attracted the attention nor found the application which
they doubtless deserve.

The sanitary disposal of sewage from buildings not in
reach of sewers, and the means for carrying away their
liquid wastes, must be accomplished by individual efforts
of their owners. This problem grows in importance every
day. That it is not so beset with difficulties as would
appear at first sight, the following considerations will
show.

In rural districts, the scattered farm and country houses
having plenty of land about them should not find it
difficult to dispose of the sewage and liquid waste matters
in an innocuous manner. Nevertheless, we encounter in
the majority of cases the use of the objectionable cesspool
and of the equally pernicious privy vault.

Cesspools and Privy Vaults to be Condemned. — Why is
it that in this enlightened age such relics of barbarism
are tolerated in civilized communities? Why is it that
for one house in the country, having a judicious and
sanitary method of sewage disposal, we find fifty houses

provided with the most primitive and obnoxious arrangements?

I believe one reason for it may be found in the universal and deplorable lack of interest in sanitary matters and in the imperfect appreciation of health laws. But why is it that the common cesspool and the vault are so severely condemned by sanitarians?

It is an axiom of sanitary science that organic waste matters, and in particular household liquid wastes, should not be stored up for any length of time near habitations, because if so dealt with they soon undergo a dangerous process of putrefaction and thereby become the cause of the pollution of the atmosphere and of the soil.

The ordinary cesspool in the country is never built tight. It is a large hole, carelessly excavated in the ground, the sides being walled with loose stones and the top covered with a loosely set flagstone or with a number of wooden boards. The liquids emptied into it soak away into the ground and very often contaminate the farmer's well. After years of use its openings become clogged with grease and the cesspool overflows on the surface or else backs up into the cellar.

The ordinary privy vault is not any better, because it is rarely water-tight; it constitutes a nuisance to sight and smell besides attracting numerous flies, as well as mosquitoes, which become quite frequently the propagators of serious disease. It always contaminates the soil and is very difficult to clean. Both the cesspool and the vault are relics of bygone times when no thought was given to sanitation.

Burning of Sewage Impracticable. — One should not, however, condemn a thing without being prepared to offer a proper remedy, and this I shall proceed to do in the following pages. One suggested solution of the problem, frequently encountered when the subject is discussed by laymen, is to get rid of the sewage by burning it. But such a plan should be at once dismissed, for it is entirely impractical and arises from a gross misconception of the problem and a confusion in the layman's mind of sewage with night soil and garbage, which latter can be destroyed by fire, while sewage is a liquid containing only two volumes of solids in 1000 parts of sewage, and hence cannot be burnt.

Waste Disposal for Small Farmhouses having no Water Service. — Let us consider first the *smaller farmhouses*, which do not have a water service or any of the modern plumbing conveniences. It would seem that in such a case no serious difficulties could exist. A clean and well-kept outside *earth closet* forms the best available substitute for the common privy. It is surprising to find how little farmers in the country know of the earth closet. This sanitary contrivance has a vault with water-tight floor, placed not much, if any, below the grade level. On the floor is set a box holding garden earth or ashes. A hinged door at the back permits of the frequent removal of the box. To facilitate this, the box is set on wheels, and in this way it is easily drawn away to the fields or to the manure pit. Either dry earth or finely sifted ashes are used. The contents of the box are often dried in a shed and then used over and over again. Much missionary work can be done in the way of educating farmers and country people to the use and appreciation of the earth closet.

For the *slop water*, which every such house, no matter

how small, necessarily has, there is an equally simple and inexpensive system of disposal. House slops should not be thrown, day after day, on the same spot in front of the back or kitchen door, for this would soon cause the soil to become saturated and a wet, mucky and offensive spot would result; the sun's rays would cause foul odors to arise, while in other cases the foul liquid may soak away to reach the nearby well, or the rain water in the cistern if the same happens not to be cemented.

In such houses the greasy kitchen water, and the laundry and chamber slops should be gotten rid of promptly by some simple form of disposal which may be accomplished in several ways. A tightly covered slop barrel or bucket may be kept on wheels near the house; the slops are carried and dumped into it, and the barrel should be carried away to the garden when full. Here the contents are buried in the ground, taking care always to select some new spot or place of disposal so that the oversaturation of the soil may be prevented.

In other cases a slop hopper is used and a small drainpipe, with tight joints, conveys the slop water from the house to a place in the vegetable garden or in the pasture where it may be disposed of safely (see Figs. 72 and 73). Such a drainpipe conveying liquid sewage should never discharge into a road gutter or into stagnant ditches, or into shallow ponds. In some cases we find an open perforated trough or gutter used, or else an open surface gutter is built with bricks laid with open-spaced joints.

Sometimes a grease trap is placed directly outside the house to intercept the kitchen grease and thereby to prevent the drain from becoming stopped up. In still other cases the drain from the house may be carried to a small cesspool. This should be built perfectly water-tight; it should be emptied frequently and the liquid used either in the vegetable or fruit garden, or in the hothouse or conservatory. Many experienced gardeners testify to the manurial value of chamber slops. The cesspool should be ventilated, if it is practical to do so, but it should never be built under or close to a house, and should, as a rule, not be closer than one hundred feet to a well or cistern.

If the house stands on high ground, sloping away rapidly, a cesspool may be built in the lowest corner of the lot and far away from the house, provided its top is not higher than the level of the bottom of the well. Under these conditions the tight bottom in the cesspool may be omitted to provide for soakage, but the arrangement should nevertheless always be considered in the nature of a make-shift.

It should be the rule never to empty house slops into the privy vault or into the earth closet. The well from which the household derives its water supply for drinking purposes should be most carefully protected from impurities and from surface drainage. Where a rainwater cistern is used this should also be kept clean and its overflow should never connect with any drain or sewer from the house. Simple sanitary arrangements, similar to the above, should be provided in the case of temporary or permanent summer camps.

Waste Disposal for Larger Farmhouses with Water Service. — In many sections of the country the *larger farmhouses* are now being provided with plumbing fixtures, such as a kitchen sink, a set of wash tubs, a bath tub, and possibly an indoor water-closet. This modern tendency of using plumbing appliances in farmhouses, which was already spoken of, is much to be encouraged, for it is the surest sign of improvement in the sanitary conditions on the farm.

In this case, too, the difficulties in the proper and safe disposal of the sewage, particularly if no water-closet is used, are not insurmountable. The ordinary leaching cesspool should never be used, but the house sewage should be conveyed by a tight drain to a flush tank with intercepting or scum chamber. The latter retains the solids and the grease, and a bent overflow pipe conveys the liquid sewage to the second chamber, from which it is discharged intermittently onto an irrigation field.

The irrigation field should not be located too close to a house, nor too close to the well or spring which furnishes the water supply. The sewage may be disposed of either on the surface of cultivated land in simple trenches following the contours of the land, or it may be disposed of by a subsurface irrigation system, consisting of a network of small tiles laid with open joints close to the surface. Experience shows that both systems continue to work in winter because the temperature of fresh sewage is sufficient to melt snow and ice in the sewage field.

Sewage Disposal for Large Country Houses. — Larger country houses, institutions outside the city limits, and summer resorts are more lavishly fitted up with a complete water-service system, indoor water-closets and other modern conveniences. They use proportionately a great deal more water, which in turn after becoming befouled, requires a somewhat more elaborate and complete treatment.

Water Pollution by Sewage of Summer Resorts. — Many summer hotels and large farm boarding houses, located on the shores of lakes or on the banks of streams, pollute these water courses by discharging a large volume of sewage into them. They thus form a menace to the health of the people further downstream who may use the water for their water supply. Hence the arrangements for the disposal of the sewage for summer resorts should be rigidly inspected annually by the health authorities, and offenders in the matter of stream pollution should be punished. The entire sanitary conditions of such places often require a very careful and strict inspection, and no nuisances of any kind should be tolerated.

There is rarely any serious difficulty in dealing with the sewage problem of such large buildings, for there are a number of methods of sewage purification available for use. It should, however, be remembered that each case forms a problem in itself, and no rigid rule can be formulated to be applied indiscriminately to all cases.

Principles of Sewage Disposal. — All the available and successful systems of sewage purification have a common principle. It is now well recognized that the microbes in the soil and in the sewage play an important part in Nature's methods for the purification of sewage. The top layers of the soil constitute powerful purifying agents and this is due to the presence in them of numberless harmless, or rather helpful, bacteria. These convert the waste matters poured into the soil into their elements, which in turn serve as plant food; they prevent the putrefaction of organic matter and accomplish the destruction of all elements which are, or may become, in any way dangerous.

Biology teaches us that there are two principal classes of such useful microbes or bacteria — the anaërobic and the aërobic. These two classes have somewhat opposite tendencies and properties, for those belonging to the anaërobic group live and grow in darkness and under exclusion of air and light, while those of the aërobic group multiply and thrive only in abundance of air or oxygen. Experiments have determined that each group can be made to perform a useful function in sewage purification.

The anaërobic bacteria serve to break up the sewage and to liquefy to a certain extent the solids in the sewage; their services constitute a preliminary treatment of the sewage. Afterwards the aërobic bacteria continue the purifying process by acting upon the partly disintegrated and liquefied sewage and changing it with the aid of oxygen into harmless elements. In modern systems of sewage disposal both kinds of bacteria are utilized, and the result of their combined work is the conversion of the sewage into a clear effluent which usually can be discharged into a stream or water course without danger of polluting the same. In exceptional cases, where a high degree of purification of the sewage effluent is called for, a secondary process of purification is necessary.

Application of the Principles. — So much regarding the principles of purification. As regards the application of the principles in the so-called bacterial purification methods, we find that the partial purification work of the anaërobic bacteria is accomplished in the intercepting chamber, or the tight cesspool with submerged inlet and outlet pipes, which devices are sometimes designated by the more scientific term "septic" or "putrefaction" tank. The subsequent oxidation and nitrification of the organic matters by the aërobic bacteria are performed either in the surface or subsurface trenches of irrigation systems, or in the sand beds of intermittent sewage filters, or in the stone beds of the intermittent sewage contact beds. The latest development of the art provides continuous trickling, percolating or sprinkling filter beds for sewage purification.

The constructional features of sewage disposal works for larger country houses will be referred to again, in the next article, which describes and illustrates a number of examples taken from actual practice.

Sewage Disposal for Village Houses. — All methods outlined in the preceding are based upon the assumption that plenty of ground exists around the house. But the dwellings in many suburban places and in villages are not so fortunately located; for as these places grow in population the houses become placed so closely together as to make it necessary to give up any attempt at local disposal. The village community as a whole is then called upon to devise and carry out proper and sufficient measures for sewage removal and sewage disposal.

Sewerage Systems. — This implies the planning and construction of a system of sewers designed to receive foul wastes only. The rainfall must be taken care of separately, and rain leaders must in no case be connected with the sewers. The reason for this is that at every rainfall the volume of sewage to be treated and purified would be unduly increased, and the difficulties and the cost of a proper disposal would likewise increase. The design of such a system of sewers and sewage disposal should be intrusted to expert sanitary engineers; the work should be carried out under their constant supervision. Approved systems generally consist of small lateral or "sanitary" sewers, with a few sub-mains and one or perhaps several main sewers, which convey the sewage to some outfall or to a sewage disposal plant.

Sewage Disposal Methods. — This outfall should never be into a ditch, brook, creek, river, stream, or lake, for the crude discharge of sewage into them would inevitably lead to their pollution. In many States, the immediate discharge of the sewage into the water courses is wisely forbidden by law, and I hope the day is not far distant when every State in the Union will have enacted such statute laws to prevent the fouling of streams and water courses of whatever description. *The prevailing view that rivers and streams may be considered as the natural outlets for sewage must be corrected.* It is only in exceptional cases that the raw sewage can be permitted to be discharged into a water course having a sufficient volume and velocity of current to dispose of the sewage "by dilution."

The common methods of sewage disposal and of water purification have always appeared to me to be in many respects not only injudicious but utterly wrong. To discharge the sewage from a town into a water course, and thus to dilute the sewage may be a very convenient method for the town, but so far as the nuisance arising from the sewage is concerned the method is only a makeshift, because it merely transfers the nuisance to the stream. We find in many instances a deliberate contamination of the water intended to be used for supply purposes, and which, before being so used, must be purified at enormous expense. Unwise and impracticable as this seems to be, it is just what is at the present day being done in some communities.

Proper methods of sewage disposal for village communities and summer hotels should be devised and installed, and the State health authorities should insist upon this.

The watersheds of lakes, rivers, and other surface waters, the surroundings of springs and the water in wells should be properly protected, and in this way the problem of water supply would find a solution which, to my way of thinking, would be much more rational and correct.

It is true that the filtration of water, if properly arranged, or the treatment of water with copper sulphate on a large scale, accomplish much good, but they will not and cannot always wholly counteract the serious effects of water pollution by pathogenic germs in the sewage, for some of these germs may be left in the water even after filtration or after the copper sulphate treatment.

Typhoid Fever due to Contaminated Water Supplies. — Typhoid fever is one of the principal diseases caused by a polluted water supply. There have been recently published, for instance, somewhat startling statistics regarding the prevalence of typhoid fever in the State of New York. It is stated upon good authority that in the nine months from January to September, 1905, no less than 60,000 persons in the State have been attacked with this disease, this being at the rate of nearly 7,000 cases per month, and that the deaths in New York City alone amounted to over 500, and this although the disease is known to sanitarians to be entirely preventable. Why is there such a very high rate of sickness and comparatively high mortality rate? What *shall* be done, what *can* be done, to prevent the ravages of this terrible disease?

Pollution of Lakes and Streams. — The fact that the streams of the State, from many of which cities and towns draw their water supply, have been contaminated by the discharge of sewage into them, has been known for years, but no energetic steps have been taken to prevent the pollution. This unsanitary practice should be stopped. What has been said of streams or running water of any kind is also applicable to lakes and larger ponds. The State health authorities everywhere should institute every year, before the warm season approaches, a thorough investigation of the sanitary condition of all summer resorts, particularly as regards their water supply and their methods of sewage disposal. The importance of such work becomes apparent from the recital of a few recent instances quoted from a bulletin of the New York State Board of Health.

"In one hotel, situated on beautiful Lake George, the hotel sewage was found to be discharged into the lake *but a short distance away from the intake which supplied the hotel with water*. Another hotel at the Thousand Islands was taking its water supply from a bay in which there was little or no current, and into which the sewage from hotels accommodating hundreds of people was daily discharged. A hotel on Long Island was found where the sewage from the hotel was discharged into a series of cesspools placed *along the front of the building directly under the windows of the rooms occupied by guests*. The cesspools were but scantily covered and the ground was saturated with filth."

State Laws to be enacted to prevent Water Contamination. — Where improvements in the water supply or the mode of sewage disposal are imperatively called for, the State health authorities should insist that they be carried out. The general public can do much in the way of assisting the work of the authorities by insisting that the sanitary arrangements of the summer resorts which they patronize be free from danger to health, and particularly that a pure water supply be provided, as is no doubt the case in many of the better class hotels and boarding houses. What is true of New York State applies equally to other States. If their public health laws are not at present framed so as to provide the required authority to do this, it is high time that the laws be changed correspondingly.

Sewage Purification the Remedy for Water Pollution. — Our brooks, our rivers, and our lakes should be kept free from pollution. No riparian owner can claim the right to discharge his sewage, without a previous thorough purification, into any stream or lake. There is a moral side to the question of water pollution which every citizen is bound to, or should be made to, respect. As pointed out heretofore, many summer resorts, some country houses, and a number of villages and towns, are the offenders. The science of sewage disposal has entered a stage where it is feasible to cope with the difficulties of the problem by providing, designing, and constructing suitable and efficient sewage purification systems.

BACTERIAL METHODS OF SEWAGE DISPOSAL FOR FARMHOUSES, COUNTRY ESTATES, AND SUMMER RESORTS

Old-fashioned and defective methods of disposing of the liquid and semi-liquid wastes from isolated households by means of privy vaults and cesspools have been discussed in the previous articles. The evils connected with privy vaults are now too well known. Cesspools should always be considered bad and unsanitary arrangements. This statement is true whether they are the so-called open or "leaching" cesspools, or else of the type known as "tight" cesspools, which, to overcome the difficulty and annoyance of frequent emptying, are generally provided with an overflow pipe, permitting the foul liquid to escape into an open ditch or into some water course.

Land Treatment. — For many years sanitarians and engineers have again and again pointed out better, more rational, and safer methods of sewage disposal in connection with the "water-carriage system," such as land treatment by means of surface irrigation, or by intermittent downward filtration. In the former system sewage acts as an irrigant of the soil, and large areas of land are required to prevent the oversaturation of the soil. In the latter system the land, consisting usually of specially prepared and well-underdrained beds, is flooded intermittently with much larger volumes of sewage than in surface irrigation.

Subsurface Irrigation. — A much-used and quite successful modification of the latter system is the disposal of sewage, known as "subsurface irrigation," which is sometimes designated as the "Waring" system, because of its having been introduced and ably advocated by the late Colonel George E. Waring.

A complete description of this method of sewage disposal in its application to country dwellings is given by the author in his books "The Disposal of Household Waste" and "Sanitary Engineering of Buildings," and the reader will find in the latter book several examples of the system with illustrations. Hence it does not seem necessary to refer to it at length in this chapter, but later on several instances will be illustrated in which the subsurface irrigation system is combined with one or the other of the artificial bacterial methods of disposal.

The method termed "disposal by subsurface irrigation"

has become well-known and is extensively applied in the case of isolated country houses. A number of contracting engineering firms make a specialty of putting in such plants for owners, generally for a fixed lump sum, and the fact should be noted that not all the systems so put in have been successful in the long run. This may have been partly due to the fact that the contractors were not sufficiently conversant with the practice of the art; it was also due, without doubt, in many cases to the contractors' desire to make large profits from a job.

The writer holds to the view expressed by him repeatedly many years ago, that it is better to have this kind of work done on a percentage basis, or on the "cost plus a fixed sum" plan. At the same time he recognizes the fact that it seems almost impossible to combat the average architect's or owner's idea that a sewage disposal system can be bought at a fixed price, much the same as an automobile or a steam yacht. These remarks apply equally to the more recent bacteriological sewage disposal plants. There is obviously quite a difference between a cheap system put in by contract which may "just do the work" and a system planned with much study and care, based upon years of practical experience in this branch of engineering, and carried out with skill and judgment with a view to its working properly for a great many years to come. There are, however, but few people sufficiently enlightened to appreciate the difference, except when it is brought home to them forcibly, as when the cheaply installed system begins to give trouble, which it generally does after one or two seasons' use.

Wherever suitable soil and suitably located land are available the methods of disposal mentioned, which are also in a sense biological methods, because the purification going on in them is caused by bacteria in the sewage and in the soil, are well adapted and generally, with some supervision given to them, have proved quite satisfactory, not only in preventing a sewage nuisance but also in yielding an effluent free from objection.

Disposal of Sewage by Dilution. — In many cases, however, sufficient land is not available, and for larger buildings, such as summer hotels and institutions, the subsurface disposal system becomes quite expensive. Another method of disposal, known as the "dilution" method, in which the crude or unpurified sewage is permitted to flow into a large water course, or into a harbor or the sea, cannot often be practiced, simple and economical as it is. Under conditions such as described the more recent bacterial or biological methods of disposal are more adapted and promise to become universally successful.

Artificial Bacterial Sewage Treatment. — These artificial bacterial methods comprise septic tanks, cultivation or upward filtration tanks, bacterial contact beds and trickling sewage filters, and this article is intended to explain briefly the principles of the methods, and to illustrate their practical application by several examples. So much has been said and written of late in particular about the "septic tank method of sewage disposal" that it would seem desirable to point out the real object of the septic tank, to define its general usefulness, but at the same time to establish clearly its limitations.

Composition of Sewage. — By way of introduction, attention should be called to the difference in composition which exists in nearly all instances between town sewage and the sewage from isolated country houses, for this has an important bearing upon the method of treatment adopted.

Sewage, in general, is a complex liquid containing organic as well as inorganic matters both in solution and in suspension, and varying in its volume and in its chemical composition at all hours of the day and night. Where household sewage proper is to be dealt with, the variations in the sewage are much less than where sewage includes manufacturing or trade wastes, and where the rainfall is admitted. The sewage from isolated houses consists of slop water, mixed with the liquid and semi-liquid excretions from men and animals. The slop water is composed of kitchen wash water, suds from the laundry, waste water from personal ablutions, dirty water from floor scrubbing and general scouring, and the drainage water from stables.

This befouled liquid, which always contains a large number of bacteria, requires purification no matter whether water-closets are connected to the house drains or not. It should be mentioned here that, whatever the sewage disposal system may be, the rain water from country houses should never be connected with the sewer carrying the foul sewage.

Town sewage, on the other hand, is apt to be more diluted, particularly after rain storms; it is also often diluted by inleakage of subsoil water, and it is nearly always mixed with trade and manufacturing wastes. It contains a good deal of inorganic or mineral matter, road detritus, silt and street sweepings. Such town sewage generally arrives at the outfall or the disposal works in a less fresh condition and more broken up.

While it always requires a basket or cage screen to intercept the coarser suspended matters and a grit chamber to eliminate the mineral matter, preliminary treatment, as described further on, may often be dispensed with for the reason that breaking up and maceration occur while the sewage matter is passing through the sewer. In very long main sewers some septic action also takes place which has a tendency to liquefy suspended matters.

On the other hand, the sewage from isolated buildings, when delivered at disposal works, is generally *fresh* sewage, because the run of the house sewer is a short one; the sewage is apt to be more concentrated and forms a good deal more scum or grease in the tanks; it also contains more solid organic matters in suspension. Hence such sewage absolutely requires some kind of preliminary treatment in order to liquefy the solids and suspended impurities, or at least to hold back the sludge which otherwise is sure to give considerable trouble in the subsequent purification methods.

At the outset I should state that, since the researches of the Massachusetts State Board of Health, and those of Warrington, Schloesing, Muntz, Dibdin, Colonel Waring and others, we know that all sewage disposal processes, except the purely mechanical straining processes, are *natural*, since they are accomplished by natural agencies, to which we only render assistance by proper forms. All forms of land disposal and purification are now known to be based upon bacterial action. In those systems which generally go under the name of biological or bacteriological methods, we have an *artificial* treatment only in so far as we provide suitable culture or growing places for the bacteria and thereby increase and promote bacterial action. The purification process itself is a natural process. All we do, therefore, is to assist nature, and if we do this wisely and in accordance with facts known through scientific research we shall succeed better than if we attempt to control nature, as for instance is done in some of the chemical treatments of sewage.

Principle of Biological Sewage Disposal Methods. — The biological methods of sewage purification are based upon the fact that all sewage contains numberless bacteria, most of which are not only harmless but useful in acting upon the sewage matters in suspension as well as in solution. The really harmful bacteria, the pathogenic or disease germs, are usually small in number; sometimes they are absent altogether, and they occur only where the bowel or other discharges from patients ill from zymotic disease, such as typhoid fever or cholera, are permitted to go without disinfection or sterilization into the sewer.

Aërobic and Anaërobic Bacteria. — There are two classes of useful bacteria, namely the anaërobic bacteria, which live and grow only in absence of light and air, and the aërobic bacteria, which on the contrary require the oxygen of the air to live and perform their functions. The anaërobic bacteria act upon the organic matters in suspension in the sewage by liquefying and gasifying the same; the aërobic bacteria act upon the organic matters in solution and assist in the processes of oxidation and nitrification.

Two Stages of Purification. — In the case of sewage from isolated buildings, sewage treatment comprises two successive stages, namely:

(1) A preliminary process for the removal of the polluting matters in suspension (by septic or cultivation tanks, or sometimes by coarse contact filter beds);

(2) A purification process for the oxidation and nitrification of organic matters in solution (by bacterial contact beds, percolating filters or by land treatment).

In order to attain success the order of these two processes should never be reversed. Sometimes, in cases where a very high degree of purification is demanded, a third or finishing process is used, consisting of either land treatment, mechanical filtration or subsidence in settling basins.

Formerly, the first part of the treatment, consisting of the elimination of the suspended matters, was accomplished by sedimentation or by subsidence in tanks, through which the inflowing sewage passed with a very small velocity of flow. Attempts were also made to roughly strain sewage in coarse filters, or to accelerate deposition by adding suitable chemicals, thus converting a mechanical process into a chemical precipitating process.

Mouras' Automatic Sewage Tank. — As early as 1881, it had been recognized by a Frenchman named Mouras that suspended organic matters became liquefied in a closed tank. He designed what he called an "automatic sewage tank," intended for the purpose. Still more recently scientific investigators recognized that the partial liquefaction or destruction of the sewage sludge was accomplished by bacterial action.

Septic and Cultivation Tanks. — In 1890 an English engineer, Scott-Moncrieff, devised a liquefying tank which he called a "cultivation tank," to which I shall refer again later on. Mr. Cameron, of Exeter, England, introduced in 1896 a putrefaction tank which he designated a "septic tank," claiming that it was far better to encourage the growth of useful microbes in tanks than to kill or destroy them by chemicals. Simultaneously with him, Mr. W. J. Dibdin, an English chemist, came to the same conclusion, and argued that the sterilizing action of chemicals or disinfectants interfered with the second purification stage, viz. the nitrification and oxidation of the sewage effluent.

Other sanitary experts, among them Colonel Waring, Lowcock, Stoddart, and Colonel Ducat worked on similar lines. In his many executed schemes for sewage disposal by subsurface irrigation in absorption tiles, dating as far back as 1880, Colonel Waring always used intercepting tanks for the retention of the solid matter in sewage, and the design of these was practically equivalent to that of the modern septic tank, i.e. inlet and outlet pipes were suitably submerged, and the tank was kept dark and arched over, so that, in my judgment, he anticipated the septic tank of both Mr. Mouras and Mr. Cameron.

Septic or Scum Tanks. — The modern septic tank, sometimes called a "scum tank" or a "putrefaction tank," consists essentially of a water-tight chamber of suitable capacity, through which the sewage flows slowly and nearly continuously as it is delivered at the outfall, the inlets and outlets being both submerged to prevent an undue disturbance of the surface or floating scum. It differs from the "cultivation tank" and from the coarse bacteria beds in not having any material, such as broken stones, placed in it to furnish suitable surfaces for the growth or cultivation of bacteria, but both the septic and the cultivation tanks are similar as regards the anaërobic conditions maintained in them, and as regards their principal function, which is the separation and liquefaction of a part of the suspended impurities in the sewage.

Cultivation or Upward Filtration Tank. — The cultivation tank, as first designed by Scott-Moncrieff, consisted of a water-tight chamber of suitable size, with a smaller separate inlet chamber, the two being connected at the bottom by a suitable channel, covered by a grating or by perforated plates. On top of these, large broken stones are placed in the tank. The sewage passes downwards through the inlet chamber and thence upwards through the grating into the cultivation tank. There is accordingly an almost continuous slow upward flow of sewage through the tank, and the liquid escapes from the latter at the top by means of an overflow pipe located at the normal water level. The object of the filling material is to increase anaërobic conditions by "affording plenty of resting places for bacteria."

Some engineers hold the view that this upward filtration of sewage through a tank filled with a material the surfaces of which offer a good medium for the cultivation of bacterial organisms, is a very efficient means for promoting or increasing bacterial action. In both the cultivation and the septic tanks the sewage is brought into a condition in which it is more quickly acted upon in the subsequent treatment.

Septic and Cultivation Tank Effluents. — While the liquid effluents from septic and from cultivation tanks contain but little suspended organic matter, they are highly charged with putrescible matters in solution, and the liquid gives off bad odors, particularly on warm or damp days. It cannot be sufficiently emphasized that the septic

tank process is only a *preliminary process* of sewage treatment, that the effluents from septic tanks are neither clarified
nor purified, that they contain all the organic dissolved
matters which are the chief cause of the contamination of
lakes and streams, and that a further purification is in
most cases absolutely required.

Preliminary sewage treatment has in some cases been
accomplished under aërobic conditions by means of coarse
contact filter beds intended to arrest and liquefy suspended
solids. Such coarse sewage beds are, however, very apt
to clog with fiber, lint, disintegrated paper and other
suspended matters, and it is now considered a better practice to remove the suspended solids in either scum or septic
tanks, or in cultivation tanks, or by means of a combination
of both.

For disposal plants for isolated buildings a grit chamber
to arrest mineral suspended matters is not usually required,
because very little, if any, road detritus or inorganic matter
finds its way into the house drain.

Work of the Septic Tank. — The bacterial action in septic
tanks is more satisfactory if the sewage is of a uniform
character and if it is concentrated rather than diluted;
warm weather increases the action of the anaërobic bacteria.
No septic tank shows good results when first put in operation; it is necessary that the process of cultivating the
anaërobic bacteria be carried on for some weeks before
the liquefying process becomes efficient.

The claims that *all* the suspended impurities are liquefied and that there will be no increase in the deposit of
solids or in the scum in a septic tank have not been
realized. On an average, only from 30 to 50 per cent of
the suspended solid matters are destroyed, partly by

liquefaction and partly by changing them into gaseous form.

Size of Septic Tank. — It seems best to make the capacity of the septic tank, in the case of country houses, equal to three-fourths of the daily volume of sewage. If made smaller than this, it becomes rather a mere settling tank; if made too large, on the other hand, causing the sewage to remain too long in the septic tank, too much anaërobic action may take place, which is found to be detrimental to subsequent oxidation. But by using two septic tanks in series, or one septic tank and one cultivation tank (as shown in some of the examples), the capacity of each tank may be reduced.

Open and Covered Septic Tanks. — For isolated buildings it is preferable to use covered tanks, notwithstanding the fact that open septic tanks have been found to be quite efficient. The reasons for this are that it is thus possible to confine bad odors and to prevent a possible nuisance near a building; the sewage scum is concealed from sight; the surface of the sewage in the tank is protected from wind, rain, and snow; cold is better excluded in winter and a more uniform temperature of the sewage is maintained, but most important of all, the possible infection of food in the household by flies, which may have settled on the scum of the tank, is prevented.

An advantage of both septic and cultivation tanks is that they do not require any fall, whereas all filter or contact beds, as seen from the illustrations given hereafter, absorb at least three or four feet of fall at the sewage outfall. The flow through the tank also requires but very little attention.

Gases Generated in Septic Tanks. — The gases generated in septic tanks consist largely of sulphuretted hydrogen, marsh gas, and hydrogen; these are inflammable and on this account it has been suggested to utilize them. The volume available from a plant for an isolated building would not be sufficient to make it worth while to attempt this.

If the septic tank is located close to a building it is best to confine the gases in the tank, and in that case it may be necessary to use some caution to prevent ignition of the gases with the possible result of an explosion and injury to life. If remote from habitations, suitable vent pipes to outdoors can readily be arranged for.

Contact Filter Beds and Trickling Filters. — We must now turn our attention to the second stage of sewage treatment, which aims at the conversion of the dissolved organic matter into innocuous inorganic compounds or elements. As has already been stated, this treatment is always necessary wherever a high degree of purification of the sewage is required. It is accomplished either by land treatment or by treatment in artificial filter beds.

The action in these is largely aërobic, i.e. it is performed by those bacteria which require the presence of an abundance of oxygen for their work. The oxidation and nitrification of sewage is, therefore, not a merely chemical, but essentially a biological process.

As early as 1882, Warrington suggested the construction of artificial nitrifying beds or filters. In recent years, two forms of artificial bacteria beds have been used for the purification of sewage, namely:

(1) The contact beds, which are filled and emptied alternately; and

(2) the trickling, percolating, or sprinkling filters, over which and through which sewage is passed either intermittently or sometimes continuously.

Practical experience seems to point to the fact that in both types the process will be more effective if preceded by a preliminary treatment in a septic tank, for we then incur less danger of the clogging of the contact bed or of the trickling filter. Nevertheless, scientific authorities are not agreed on the question whether a preliminary anaërobic treatment really facilitates the subsequent oxidation and nitrification of sewage. Some claim that preliminary treatment is not required in the case of the trickling filters, but European practice seems to confirm the opposite view. The writer holds that for a purely domestic sewage from isolated houses, coming to the outfall unbroken, fresh and undiluted, preliminary treatment is essential for a thorough purification, and also to prevent trouble from the clogging of the beds, or in the subsurface irrigation disposal the clogging and stopping up of the absorption tiles.

Bacterial Contact Filter Beds. — Bacterial contact beds consist essentially of water-tight open tanks, generally built in concrete or brickwork, and filled with a coarse-grained material suitable for bacterial growth. (See the illustrated examples.) They are provided at the top with sewage inlet and with distributing troughs, and at the bottom with open-jointed drain tiles and emptying pipes closed by gate valves. The object of the filling material is obviously to expose a maximum of surface alternately to the sewage and to the air.

Contact beds are charged at regular intervals with sewage, and before doing this the outlet valves are closed. The sewage is then left standing in the beds and "in contact with" the bacteria. From this mode of operation the name of the process is derived. After some time, generally a period of two hours, the bed is emptied and is then left standing empty for oxidation and aëration; this is commonly called the "resting period," and it may extend over from two to four hours. Afterwards the regular cycle of operations begins again.

Purification Process of Contact Beds. — The process of purification going on in bacterial contact beds is somewhat complex and difficult to describe and to define. The action is partly a mechanical or straining process by which the suspended matters, which are carried over from the septic tank, are arrested. It is chiefly, however, an oxidizing process of the organic matter accomplished through the agency of the bacteria. The stones composing the filling material become covered on their surfaces with a gelatinous growth which contains the bacteria. In passing over this the liquid sewage parts with a large portion of the material held in solution, a process designated by Dr. Dunbar, Prof. Winslow, and others as "adsorption."

This is a very important and essential, but as yet little understood, part of the purifying process. The emptying and draining of the bed draws in oxygen with the air, which comes into intimate contact with the gelatinous growth, and thus the oxidation of the organic matter by the bacteria living in the same is accomplished.

The real work of the aërobic bacteria is therefore done during the so-called "resting period" of the bed. During the filling and standing full of the bed the action taking place is at present still somewhat obscure. It certainly cannot be considered as wholly aërobic as claimed by some.

We have seen that the regular cycle of operation in a contact bed is (1) filling; (2) two hours' standing full; (3) emptying and (4) four or more hours of resting. A contact bed may receive three fillings during a period of 24 hours, but it is more usual to fill a bed only twice a day, as a better degree of purification is thus attained.

Depth and Character of Filling Material in Contact Beds. — The average depth of the filling material in a contact bed is 4 feet, though some good results have been attained with depths of only 3 feet. The bed should be thoroughly aërated by allowing the air to find ready access to the interstices of the filling material. Experience has shown that it is a mistake to make the top layer of the bed of a finer material because this readily clogs up, and aëration is interfered with more or less. Sometimes aëration is accomplished by means of short earthen pipes, set vertically in the filter bed and projecting somewhat above the surface of the bed.

Any hard, broken up material, such as hard-burnt clinker, coarse, sharp gravel, granite chips or other broken stones, are suitable for the filling. Hard coal is also excellent, but experiences with coke and coke breeze, also with soft limestone, have shown that these are not so good, being porous and subject to quick disintegration. The material should be free from dust and dirt, and should be washed before use if necessary. The size of the stones best adapted for use varies from one to two inches.

Capacity of Contact Filter Beds and Mode of Filling Bed with Sewage. — The liquid capacity of a contact bed is at first about 50 per cent of the total cubical capacity, but it

soon becomes reduced to about 33 per cent, this being
caused by the settling down of the material, the growth of
organisms, the breaking down of the filling, and the introduction of solid matter. Nevertheless, such contact beds
often work many years without requiring to be refilled.

The filling with liquid sewage should be done quickly,
hence a sudden discharge in a large volume from a collecting tank, in which the sewage has been allowed to accumulate after passing through the septic tank, is preferable
to an irregular, slow flow directly from the septic tank.
The sewage is usually distributed over the top of the bed,
in wooden or preferably iron troughs, perforated with
numerous holes; sometimes, however, the bed is filled
from the bottom, and this seems to me to be preferable, as it
does away to some extent with the odor from the septic
sewage.

Automatic Appliances for Operating Contact Filter Beds.
— The operations of opening and closing the outlet valves
may be accomplished by hand, or else they are done
automatically. There are a number of patented automatic
contrivances which accomplish this, but which will not be
described in this article, for instance those of Cameron,
Adams, Merritt, Shields, and others. It is claimed that
these appliances render the sewage purification plant
independent of manual attention.

The writer's opinion is that hand operation is, in most
cases, far preferable, as it keeps the system under more
careful observation. He finds that all automatic appliances, and in particular those having movable parts or
mechanisms, are liable to get out of order, to corrode, and
then fail to work, and hence do require frequent attention.

No greater mistake can be made than to think that once

a sewage disposal system is installed it requires no oversight or some regular attention. On the contrary, every
part of the system, including the siphons or other automatic
appliances, do need occasional looking after and even
cleaning.*

A contact bed should be carefully operated and skillfully

* That the writer does not stand alone in this view is shown by the
following quotations from other engineers and sewage experts: —

FOLWELL. — "No method of treatment is entirely automatic, but all
systems need intelligent care."

PROFESSOR FLETCHER. — "Let no one imagine that such a system
can be left to run itself. The little attention and labor bestowed are
indispensable and must be given with absolute system and regularity.
Intelligent control is the necessary condition for success."

JONES AND ROECHLING. — "Mr. Cameron and other engineers may
boast of their labor-saving automatic appliances for opening and shutting
valves on sewage works, but practical workers will agree with us in hesi-
tation as to placing entire confidence in the substitution of automatic
machines for any large proportion of the manual labor. It was formerly
maintained that neither contact beds nor septic tanks required careful
superintendence, but that they could be worked by automatic machinery
and left to themselves. This was not Mr. Dibdin's view, who after years
of careful study came to the conclusion that they were delicate pieces of
mechanism which required careful and constant watching. Mr. Dibdin's
conclusions have since been confirmed by all careful experimenters."

FRANKLAND. — "Land treatment requires less skilled supervision than
contact beds."

W. J. DIBDIN. — "It is claimed for certain processes that they will
work successfully for an indefinite period without any attention whatever,
but I have not yet seen one which is left to run alone without being
watched. Why be so anxious to procure an automatic process for
purifying waste water? No method should be proposed which will work
without supervision."

"No system will run itself; human agency must constantly intervene.
else neglect will spell failure."

"The assumption that contact beds do not require careful superintendence, that they may be worked by automatic machinery and left to
themselves is altogether wrong."

managed in order to obtain a well-purified effluent and also in order to prevent a gradual undue loss of capacity in the bed, and consequent loss of efficiency.

Sprinkling or Trickling Sewage Filters. — The latest development of bacteriological purification methods is the percolating, sprinkling or trickling filter to which sewage is applied either intermittently with periods of rest, or else continuously. It consists of a rather deep filter bed, filled with a coarse material and arranged with a view of obtaining the freest possible circulation of air through the bed. The bed is composed of broken stones, which are somewhat coarser than in contact beds and which form the filter, and there are usually no side walls, except larger stones to hold the material in place (see section of trickling filter, Fig. 92). It is by far the best to have the entire filter standing free and exposed on all sides, as shown in view in Fig. 91, and the writer is fully convinced that a higher degree of purification is obtained in such filters than in those which are sunk into the ground in much the same way as contact filter beds.

In this connection it might be well to point out that the experimental trickling filters arranged by Prof. Winslow and Prof. Phelps at the Sewage Experiment Station of the Massachusetts Institute of Technology and also those upon which Prof. Dunbar experimented at the Sewage Experiment Station in Hamburg, Germany, which the writer visited in February, 1907, are constructed so as to stand entirely free on all sides. Trickling filters for the purification of larger volumes of city sewage are likewise so arranged (see illustrated article, by the writer, on the Wilmersdorf-Berlin sewage purification plant, in *Engineering News* of March, 1908).

A trickling sewage filter differs from a sewage contact bed principally in the method of applying the sewage. There are no valves to confine the sewage in the bed, and the sewage at no time stands in the bed, but it trickles through it all the time.

Methods of Operating Trickling Sewage Filters. — The sewage does not flow onto a sprinkling filter as in a contact bed, but it is sprayed or showered over it, and special devices, such as revolving or traveling sprinkler arms (see Fig. 93), or in other cases fixed brass nozzle jets or suspended troughs with perforations are used for the purpose. The depth of the filter bed should not be less than five feet, and sometimes as much as eight or even ten feet are used where the available fall permits. The entire filter bed is thoroughly aërated, and the sewage trickles through it and passes out through underdrains placed in the bottom.

The trickling or percolating filter, as it is sometimes called, is cheaper in construction than a contact bed and it is claimed by some that it accomplishes better work. The fears that such an open filter would be interfered with by frost in our winter climate do not seem to have been confirmed, if one may form a conclusion from the results of elaborate experiments made at sewage experiment stations in Boston, Mass., and in Columbus, Ohio. The writer has recently inspected during his European trip a large installation of 57 trickling filters, intended for the purification of the sewage of 600,000 persons (a part of the new suburbs of the city of Berlin), which system worked without serious interruption from ice when the outdoor temperature was at 2° Fahr. (See description in *Engineering News* of March, 1908.)

Advantages of Trickling Sewage Filters. — The chief advantages of percolating filters over contact beds are that they do not clog up so easily, that they permit a much higher rate of application of sewage, and also that their purifying efficiency is somewhat greater.

It is stated on good authority that sprinkling filter beds may treat more than twenty times the amount of sewage per unit of area than intermittent sand filter beds.

Trickling filters have not as yet been applied to the purification of domestic sewage from isolated houses, but the scheme is worth trying, and in several of the examples below a suggestion in this direction is given.

Several patented systems, using trickling filters, employ artificial means for blowing air through the filter bed for an intensified aëration and a better oxidation of the sewage, but such methods are, as a rule, unnecessarily expensive. An account of some of these artificially aërated filter beds is given by Col. Waring in his description of the sewage purification works at Homewood, Brooklyn, and at Willow Grove Park, near Philadelphia.

Subsequent Treatment. — Where a high degree of purification is required in the effluent, the liquid flowing from trickling filters may be purified further by land filtration, or by subsidence in sedimentation basins, or by treatment in sand or gravel filter beds.

Applicability of the Different Methods described. — Both the contact beds and the trickling filters may be used in double series, and the beds are then called "primary" and "secondary" beds, and the double treatment yields a satisfactory and clear effluent. This can, however, only be done practically where there is an abundant fall at the outfall. In that case the septic or scum tank may be reduced somewhat in size.

Where fall is not so plentiful, a larger septic or cultivation tank should be used, followed by a single contact bed treatment.

It may happen in some cases that there is no fall whatever; in that case the sewage must be purified by septic tank followed by treatment in a larger cultivation tank, and finally with land purification by surface or subsurface irrigation.

Wherever suitable land for natural treatment of sewage is difficult to obtain, or where the land adapted to such purposes is held at a very high or prohibitive price, the artificial bacterial purification methods may offer a successful remedy, but the systems should be judiciously planned and require skilled labor and some judgment in the management and control.

Care and Management of Sewage Disposal Plants. — Experience with the usual types of subsurface irrigation systems has shown, in some cases, that in the course of years the absorption tiles in the disposal field are apt to become choked. When this occurs the sewage field fails to act properly, becomes sewage-sodden, and in some cases a nuisance is established. This is nearly always caused by a lack of intelligent oversight and management of the disposal plant. The fact that the sewage disposal system is fitted up with an automatic siphon discharge misleads owners in causing them to think that the entire system is automatic. No greater mistake than this is possible, for *every* part of a system requires looking after.

While a disposal plant should be designed as simply as possible, so as to be operated and maintained by a gardener or one of his assistants, no sewage disposal system will run itself for any length of time, careless statements to the contrary notwithstanding. A sewage disposal plant, much like any other apparatus or machinery, requires a certain amount of periodic attention after being installed, to secure continued good results. The owner of a country house, who buys and installs an electric lighting plant, or a pumping engine, a filter, or an automobile, would, as a matter of course, expect to have the machinery of each looked after by an attendant. Why not then detail and instruct some intelligent attendant to look after the sewage disposal plant?

The causes of trouble with sewage plants are usually not far to seek. When scum and grease are permitted to accumulate to such an extent in a sewage tank that the overflow pipe carries them into the second or liquid sewage chamber, the automatic siphon discharges not only liquid sewage, but also some of the scum, and in this way the lines of tiles in the subsurface irrigation field gradually become clogged. Sometimes the siphon itself clogs up and, instead of having an intermittent siphon discharge and a vigorous flushing action, the sewage will only dribble down to the irrigation field in a small stream. This likewise leads to trouble with the tiles in the field.

In the author's own experience a disposal system generously planned as to size worked satisfactorily and without the least trouble for a period of 14 or 15 years. At this time the gardener who had charge of the system began to neglect it. Although he noticed that the siphon failed to operate properly because it had become stopped up, he let the system run on in its bad condition, with the result that at the end of the next summer season all the lines of tiles had become more or less choked. This of course necessitated the entire relaying of the disposal field.

With a view of preventing, as much as possible, the failures, annoyances and troubles described, the author has in recent years designed and installed a number of modified subsurface irrigation systems, the modification consisting in increased means for liquefying the solids in the sewage by using septic tanks in combination with either coarse filter beds or cultivation tanks. In both cases the sewage effluent is subsequently disposed of by the sub-surface irrigation system.

EXAMPLES OF SEWAGE DISPOSAL SYSTEMS FOR COUNTRY HOUSES.

The principles of modern sewage treatment for houses in the country, dwelt on in the preceding pages, will now be illustrated by a number of examples.

I. *Sewage Disposal by a Combination of a Septic Tank, a Coarse Filter Bed and Subsequent Sub-surface Irrigation.*

As shown in Fig. 74 the sewage tank consists of a combination of a septic tank with a double set of filter beds. The septic tank is built in the usual manner, oblong in plan, with submerged inlet and outlet pipes and with a dividing wall in the center, which is intended to prevent the disturbance of the scum. The outlet pipe from the septic tank divides into two branches, each of which is provided with a shear gate valve, which permits the septic sewage to be discharged into one or the other of the two coarse filter beds.

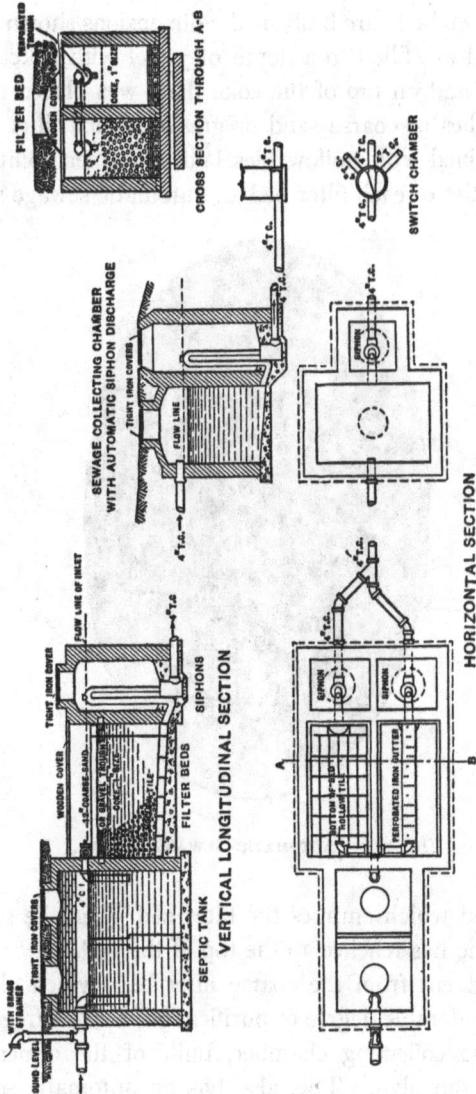

Fig. 74.—Sewage Disposal by Septic Tank, Coarse Filter Beds and Subsurface Irrigation.

The filter beds are built of the dimensions shown on the plans and are filled to a depth of 3 feet with coke of one inch size and on top of the coke there was placed a layer of 12 inches of coarse sand or gravel. Each filter bed is underdrained with hollow tiles laid with open joints, and at the outlet of each filter bed an automatic sewage siphon

FIG. 75. — AUTOMATIC SEWAGE SIPHON.

is provided which empties the filter bed when the sewage in the same has reached to the top of the bed.

The effluent from the coarse filter bed, which already has a considerable degree of purification, is discharged into the sewage collecting chamber, built of the dimensions shown in the plan. This also has an automatic sewage siphon which empties the tank intermittently and discharges the sewage by means of the switch chamber, placed at the head of the disposal field, into one or the other of several lines of irrigation tiles.

Various types of automatic siphons are used in sewage tanks, some being modifications, others improvements on the original form of Field's automatic sewage siphon. Fig. 75 shows in view a simple sewage tank siphon which operates without any moving parts and which has the

FIG. 76. — GATE OR SWITCH CHAMBER AND OTHER ACCESSORIES OF SEWAGE TANKS.

advantage that it starts siphonage very quickly. The view is taken from a photograph kindly supplied by Messrs. Waring, Chapman and Farquhar. They also supplied the view, Fig. 76, of other accessories of subsurface disposal systems, such as the gate or switch chamber in the center of Fig. 76.

Formerly round two-inch tiles were used exclusively in sewage disposal fields, but more recently the size of these

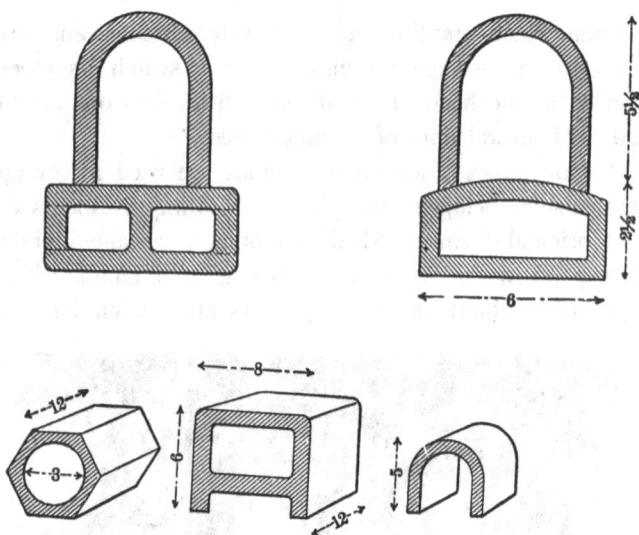

FIG. 77. — VARIOUS FORMS OF DRAIN TILES USED IN
SEWAGE DISPOSAL.

was increased to three inches, and in other cases special
forms of disposal tiles have been designed, such as those
shown in illustration, Fig. 77. The advantage is claimed
for these that they have a larger sewage carrying capacity.

The system described has worked successfully for many
years.

II. *Sewage Disposal by a Combination of Septic
Tank, Primary and Secondary Contact Filter
Beds, followed by Subsurface Irrigation.*

In this problem the drainage from a large house had to
be disposed of on a slope bordering an inland lake, the
pollution of which was prohibited by State laws, hence a
high degree of purification of the effluent was desirable.

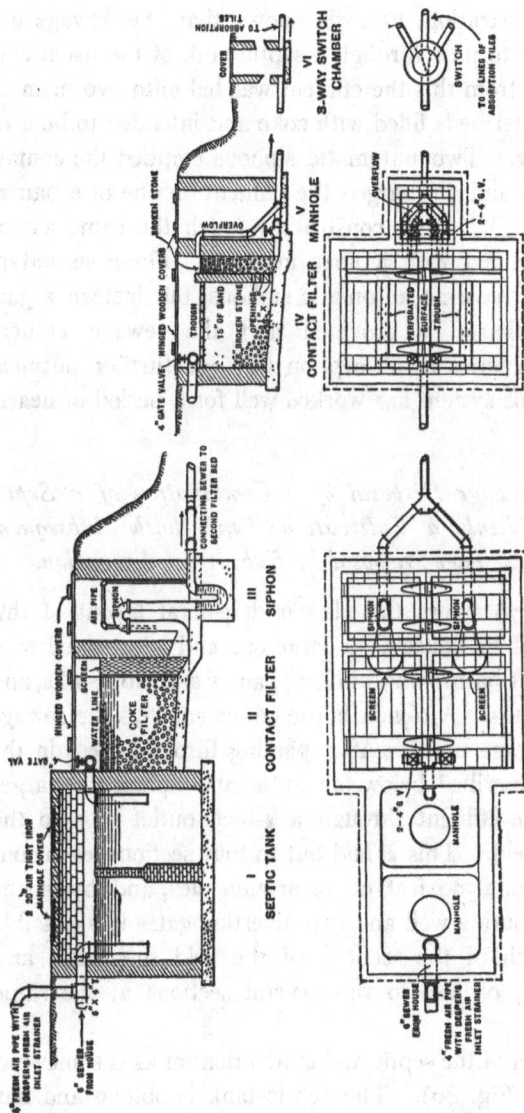

Fig. 78.—Sewage Disposal by Septic Tank, Primary and Secondary Contact Filters and Subsurface Irrigation.

The illustration, Fig. 78, shows that the sewage was first made to pass through a septic tank of the usual construction; from this the effluent was led onto two primary contact filter beds filled with coke and intended to be used alternately. Two automatic siphons emptied the contact filter beds and discharged the effluent on one of a pair of secondary filter beds constructed much the same as the first, but filled with a finer material. These secondary filter beds have no automatic siphons, but instead a gate valve discharge is provided and the sewage effluent runs to a series of absorption tiles for further purification. This system has worked well for a period of nearly ten years.

III. *Sewage Disposal by a Combination of a Septic Tank, a Cultivation Tank, with Subsequent Sewage Disposal by Subsurface Irrigation.*

Figure 79 shows the plan and general layout of this system. The house is a large one and is provided with about eight bathrooms, kitchen, pantry and slop sinks, and laundry tubs. A 6-inch house sewer conveys the sewage to the sewage tanks. After passing through these in the manner described below, an automatic siphon discharges the sewage effluent through a 4-inch outlet pipe to the disposal field. This is laid out in four sections, each containing about 500 feet of absorption tiles, and by means of a distributing well and two diverting gates it is possible to use each of the sections of the field separately and alternately, or else to run several sections at the same time.

The plan of the septic and cultivation tanks is somewhat novel (see Fig. 80). The septic tank is oblong and has

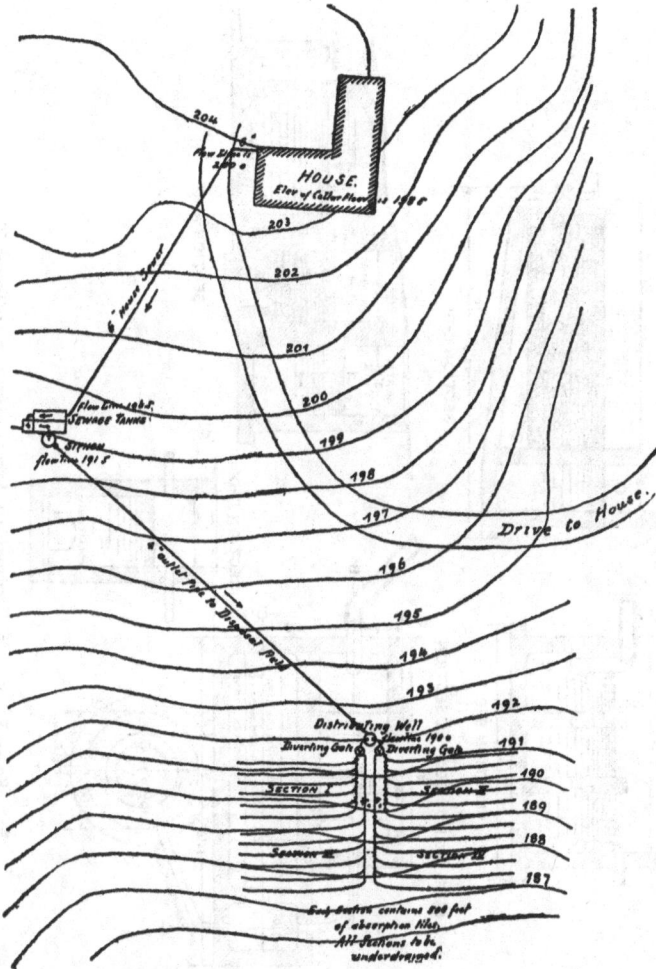

FIG. 79. — SEWAGE DISPOSAL BY SEPTIC AND CULTIVATION
TANKS FOLLOWED BY SUBSURFACE IRRIGATION.

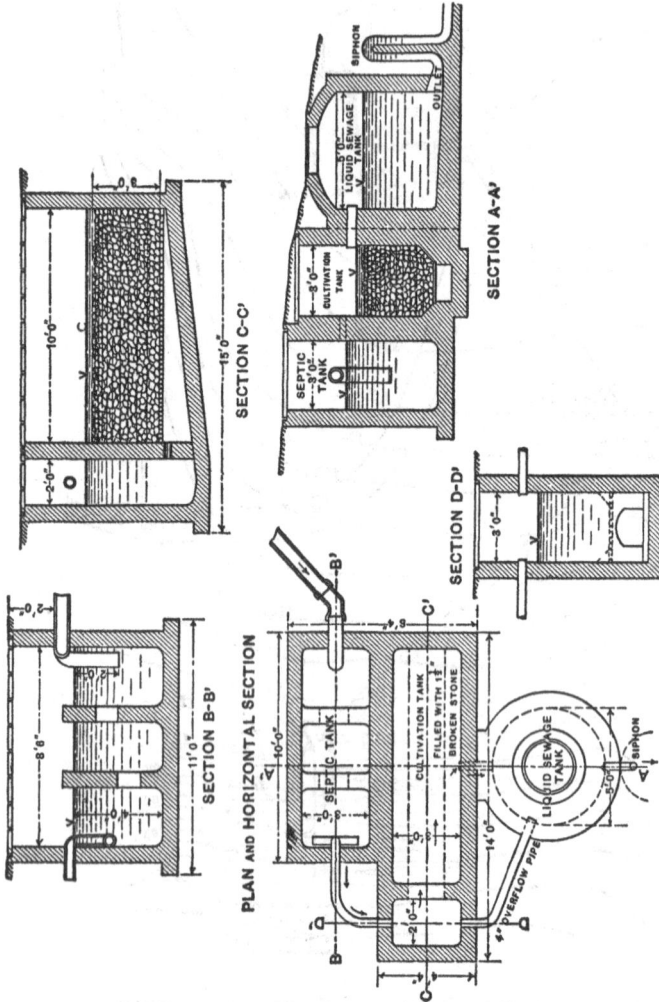

FIG. 80. — PLAN AND SECTIONS OF SEPTIC AND CULTIVATION TANKS.

two partition walls. The effluent from the tank passes through an inlet chamber into the cultivation tank. This is an oblong tank filled with broken stone, and the motion of sewage through it is upward and anaërobic action takes place while the sewage stands or flows through the cultivation tank. The effluent is gathered into a circular liquid sewage tank, the contents of which are discharged by means of the usual automatic sewage siphon.

The system described has recently been put into operation, and the author is going to watch the same with much interest, believing as he does that an upward flow of sewage, which has been partially purified in a septic tank, will secure very satisfactory results.

The disposal system, described in Example XI, and illustrated in Figs. 94 and 95, has also a cultivation tank in connection with a septic tank.

In a slightly different design for such a combination, the septic and cultivation tanks are placed one beyond the other in the same axis line. Both the septic tank and the cultivation tank are made of a capacity of 900 gallons. A sump with sluice valve and 6-inch emptying pipe is provided at the bottom of the communicating chamber between both tanks, the object being to clean at intervals the cultivation tank by reversing the flow. Such an arrangement is of course only possible where the topography of the land permits of running the emptying pipe mentioned. Instead of arching the tanks over as is customary, wooden board covers are sometimes provided on top of both tanks, which can be readily removed in order to expose the tanks fully for observation or for the removal of obstructions.

IV. *Sewage Disposal by Combination of Septic Tank with Four Contact Filter Beds, to be used alternately.*

The design of this sewage disposal plant is shown in the plan in Fig. 81 and in the several sections Figs. 82, 83 and 84. The plant is intended for a large house, having an average number of twelve occupants, the daily water consumption being 1200 gallons. The septic tank, the collect-

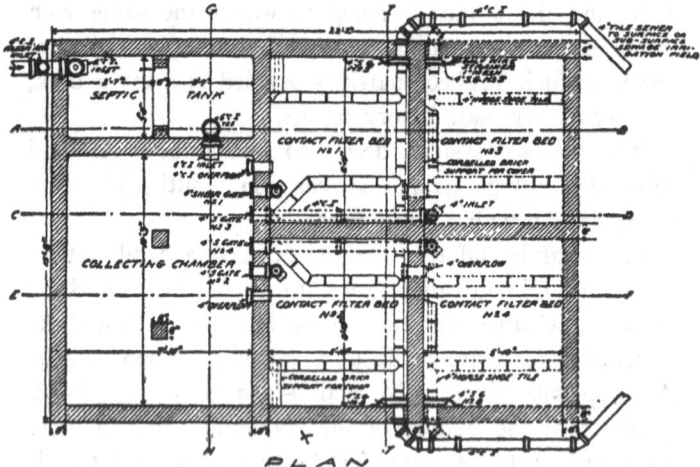

FIG. 81.— SEWAGE DISPOSAL BY SEPTIC TANK AND FOUR CONTACT FILTER BEDS.

ing tank for the septic effluent, and the four filter beds have been planned compactly together as shown. The septic tank has a capacity of 900 gallons and the collecting tank of 1200 gallons. Each contact filter bed has a net liquid capacity of 300 gallons.

This plant is intended to be *operated* entirely *by hand*

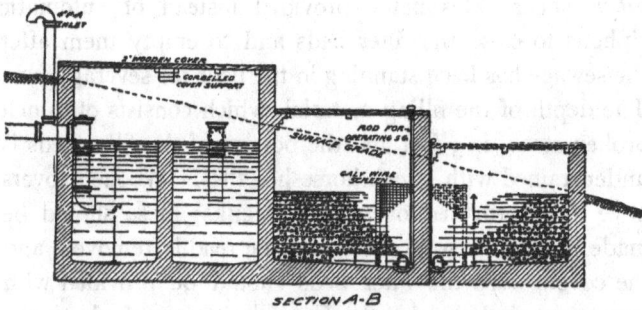

FIG. 82. — LONGITUDINAL SECTION THROUGH SEPTIC TANK AND
CONTACT BEDS.

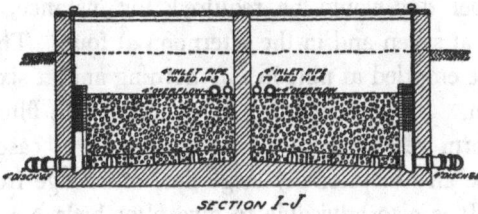

FIG. 83. — CROSS SECTION THROUGH CONTACT BEDS.

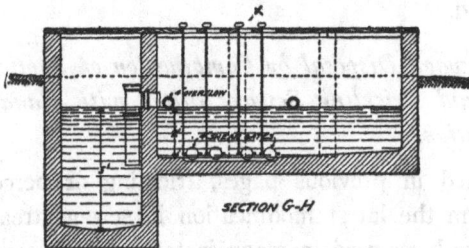

FIG. 84. — CROSS SECTION THROUGH SEPTIC TANK AND
COLLECTING OR DOSING CHAMBER.

labor, shear gates being provided instead of automatic siphons to close the filter beds and to empty them after the sewage has been standing in the beds for several hours. The depth of the filling material, which consists of 2-inch broken stone, is 3 feet and the bottom of the filter beds is underdrained with 4-inch horseshoe tiles. Wooden covers have been provided over all the tanks; these should be made in sections so that they can be readily removed, and the covers over the filter beds should be provided with numerous air holes for the free admittance of air to the contact beds.

A plant, such as illustrated, may be operated in several ways. If two contact beds only are used, two fillings per day would be required, for instance, in the morning at seven and in the afternoon at four. The beds should be emptied at nine in the morning and at six in the afternoon. In this case the third and fourth filter beds would form reserve beds to be used only in case there should be an exceptionally large flow of sewage from the house. It is also advisable to give filter beds No. 1 and No. 2 a rest of several weeks after they have been used for a month, and then to use filter beds No. 3 and No. 4 for a like period.

V. *Sewage Disposal by Combination of Septic Tank and Trickling Sewage Filter, with Hand Operation.*

As stated in previous pages, trickling or percolating filters form the latest modification in sewage treatment. They have been used in many instances on a large scale for the purification of the sewage of towns, and likewise for sewage disposal for large institutions, but no instance is on record, to the author's knowledge, where a trickling filter has been used for the purification of the sewage from a country house. The scheme, however, appears to be perfectly feasible and is certainly worth trying. In the case of trickling filters it is unnecessary to provide water-tight basins, such as are used with contact beds, therefore the construction becomes somewhat cheaper. The chief requirement is that the filter be well aërated, and for this reason it should be constructed so as to stand entirely free as shown in the section of Fig. 85 (see also Figs. 91 and 92). It may, however, become necessary to provide a simple roof over it to exclude the rain water, and also a plain enclosing structure to hide the filter from sight.

The construction should be such as to permit the spreading of the liquid sewage as uniformly as possible over the entire area of the filter. The distribution of sewage may be effected from a series of fixed nozzles, or else open gutters or suspended troughs may be provided, as shown. These should have a large number of holes. In this way the sewage is brought onto the filter in the form of a fine spray and trickles slowly downward along the broken stones which compose the filter without at any point forming a continuous stream of sewage. The bottom of the filter should be underdrained and an outlet should be provided for the purified sewage effluent. Sometimes special aërating pipes are put through the body of the filter as shown in Fig. 92.

A trickling filter should not be used without the sewage having received a preliminary treatment in a septic tank (not shown in the illustration). The construction of the latter would be similar to that of the examples previously shown and the effluent should be gathered in a collecting

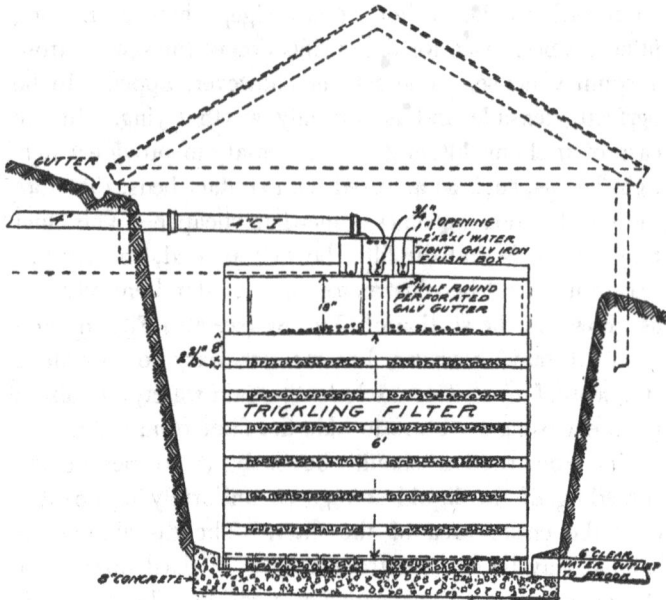

SECTION E-F

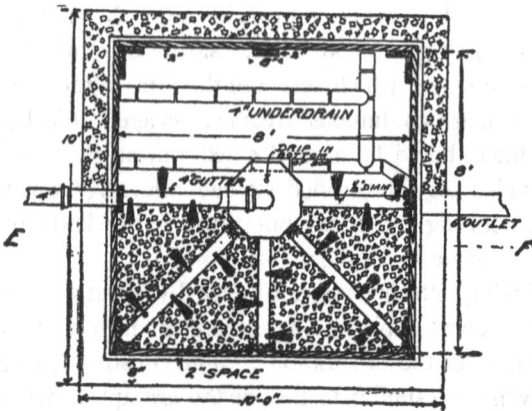

FIG. 85.—PLAN AND ELEVATION OF TRICKLING FILTER.

chamber provided at its lower outlet with a gate valve operated by hand.

In many cases the layout of the land will be such as to render it impossible to have a gravity discharge from the septic tank and from the trickling filter. In this case the trickling filter should be placed on a higher elevation than the collecting chamber and the sewage should be pumped up automatically by means of a submerged centrifugal pump, operated by an electric motor. In other respects the system would not differ materially from that shown in the example.

VI. *Sewage Disposal by Septic Tank for a Large Building or a Group of Buildings Located either on the Shore of a River which is not used for Drinking Purposes, or on a Tidal Estuary which does not contain Oyster Beds, or near the Ocean.*

Under the several conditions mentioned, a high degree of purification of the sewage effluent is not ordinarily required. But it is at all times necessary to prevent a nuisance to sight or smell at the point where the sewage is discharged. A simple preliminary treatment of the sewage may therefore be found sufficient, and for such cases the septic tank treatment is eminently adapted. If the building has an average population of 200 persons, and a daily water consumption per person of 60 gallons, the resulting volume of sewage in 24 hours amounts to 12,000 United States gallons. Rain water should, of course, be rigidly excluded from the works. The capacity of the septic tank should be made equal to three-fourths the daily consumption, i.e. it should hold 9000 gallons of

sewage.   The septic tank is best built of an oblong shape,
as shown in plan and section in Fig. 86, and its principal
dimensions would be approximately as follows: —

Total length inside,   22 feet 6 inches,
Width inside,   6 feet 0 inches,
Depth below water level,   9 feet 0 inches.

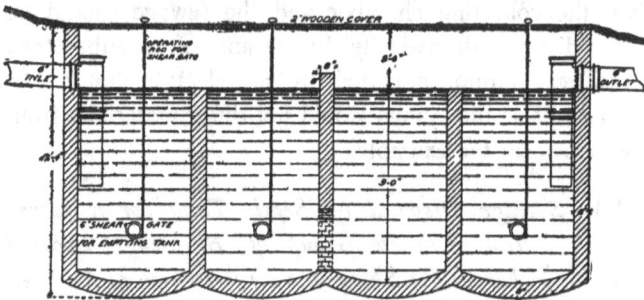

SECTION

PRELIMINARY TREATMENT IN
SEPTIC TANK BY ANAEROBIC ACTION
CAPACITY=9000 U.S.GALLS = ¾ DAILY CONSUMPTION
SCALE

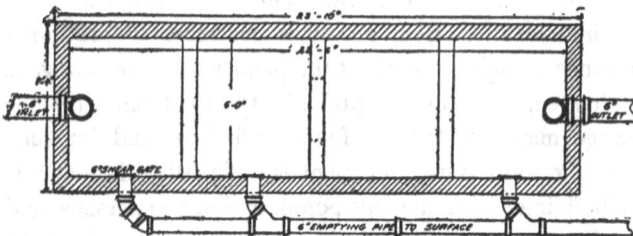

PLAN

FIG. 86. — SEWAGE DISPOSAL BY SEPTIC TANK.

The depth of excavation for the tank would be about
12 feet.   The tank could be built of brick or stone masonry,
or else of reinforced concrete steel construction.   The

inlet and outlet pipes should be suitably submerged in order not to have any disturbance of the surface scum. To further assist deposition in the tank, three cross walls are built in the same, the middle one being 8 inches higher than the water level, and having a large opening near the bottom, thus forcing the sewage to pass through the tank in a circuitous route. Shear gates are provided near, but not quite at, the bottom of the tank in each of the compartments to permit of the occasional emptying of the liquid sewage, whenever desired, without disturbing the bottom sludge.

In the normal operation of the septic tank, the gates stand closed tight. The effluent may either pass out to the water course by a tight sewer outlet pipe, or run into a specially excavated trench, about two feet in depth and filled with very coarse broken stone.

VII. *Sewage Disposal for a Large Building Located on the Sloping Shore of an Inland Lake.*

In this case no unpurified sewage should be permitted to flow into the lake, but a high degree of purification is required in view of the possibility of a water supply being taken from the lake. The water of the lake as well as its shores should also be kept pure and undefiled, in order not to prevent the use of the lake for bathing, boating, or fishing purposes; in fact any sewage nuisance must be absolutely prevented; it is further desirable that the sewage disposal plant be out of sight or made as inconspicuous as possible.

The problem cannot be solved by using only a septic tank and discharging its effluent directly into the lake, nor can such effluent be sufficiently purified by merely running it through an underground trench filled with broken stone and covered over at the top, as suggested in the previous example. Some further system of purification is absolutely required.

In cases like these it is advisable to provide for a preliminary treatment and liquefaction of a part of the suspended matters in a septic tank, which may be built as already described in example VI and in Fig. 86. As an alternative arrangement, a combined septic and cultivation tank, as shown in Fig. 80, but of larger dimensions, may be used. Adopting the second scheme, it is possible to reduce the capacity of the septic tank proper to one half of the former size, or 4500 United States gallons; the liquid contents of the cultivation tank are also made equal to 4500 gallons.

The dimensions of the septic tank would be 11 feet 3 inches long, 6 feet wide, and 9 feet depth of sewage; the cultivation tank would be made 10 feet wide, 20 feet long, and the depth of the anaërobic bed in the same is made 4 feet. The channel under the iron gratings through which the sewage flows upward is made 3 feet wide, 3 feet deep at the lower end and 1 foot 6 inches deep at the upper end. Provision is made by a shear gate at the lowest point of the chamber for the occasional removal of the sludge accumulating in the same. Both tanks may be covered in the simplest manner with wooden board covers, or else a wooden house with a light roof may be built over both tanks.

After flowing upward through the cultivation tank the effluent passes out through the outlet pipe, and is now in a proper condition for a more complete purification by oxidation and nitrification. This may be accomplished in

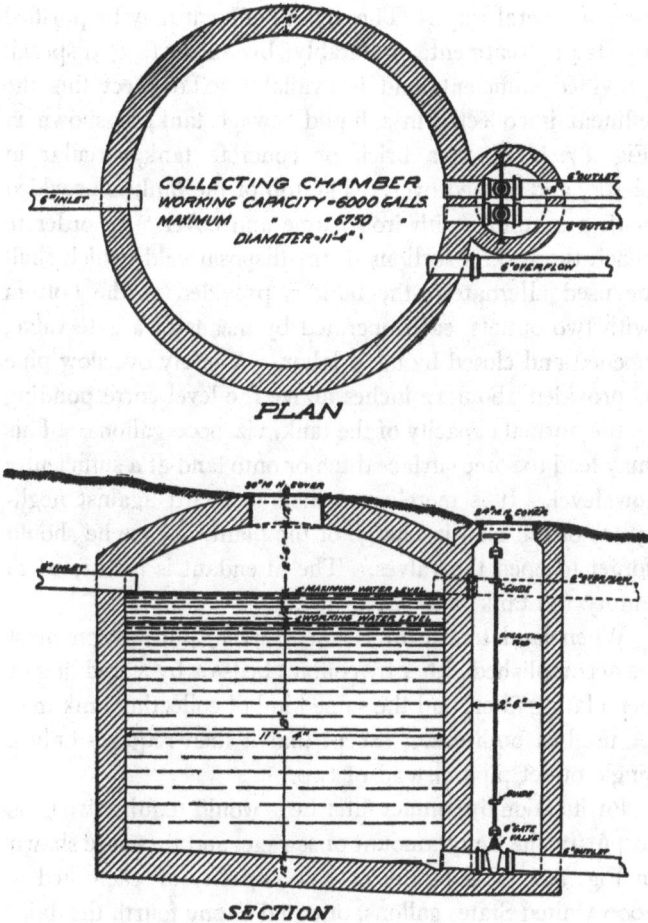

PLAN

SECTION

COLLECTING CHAMBER FOR LIQUID SEWAGE.
WORKING CAPACITY = 6000 U.S GALLS
= ½ DAILY CONSUMPTION.

SCALE:

FIG. 87. — SEWAGE DISPOSAL BY SEPTIC TANK, COLLECTING OR
DOSING CHAMBER AND BY CONTACT FILTER BEDS.

one of several ways. The sewage effluent may be purified by land treatment, preferably by subsurface disposal, provided sufficient land is available. To effect this the effluent is collected in a liquid sewage tank, as shown in Fig. 87. This is a brick or concrete tank, circular in shape, and arched over. The top of the tank is provided with a manhole with iron frame and cover. In order to reach the several sections of the disposal field, which shall be used alternately, the tank is provided at the bottom with two outlets, each operated by means of a gate valve, opened and closed by hand labor. A safety overflow pipe is provided about 12 inches above the level corresponding to the normal capacity of the tank, viz. 6000 gallons. This may lead to some surface ditch or onto land at a sufficiently low level. It is merely provided to guard against negligence of the man in charge of the plant, in case he should forget to open the valves. The attendant is instructed to empty the tank twice each day.

When land treatment is not feasible, purification must be accomplished either by contact beds or by a trickling or percolating filter, and the same kind of collecting tank may be used in both cases, except that it then requires only a single outlet pipe instead of two.

Purification by contact filter beds would require two beds to purify the daily amount of sewage, and these are shown in Fig. 88. The liquid or net capacity of each bed is 3000 United States gallons, or equal to one fourth the daily volume of sewage. The two beds occupy an area of about 28 feet square. Each bed is filled to an average depth of 4 feet with broken stones of $1\frac{1}{2}$ to 2 inch size. The bottom of the bed is suitably drained by means of 6-inch horseshoe drain tiles, with 4-inch branches. The effluent pipe is

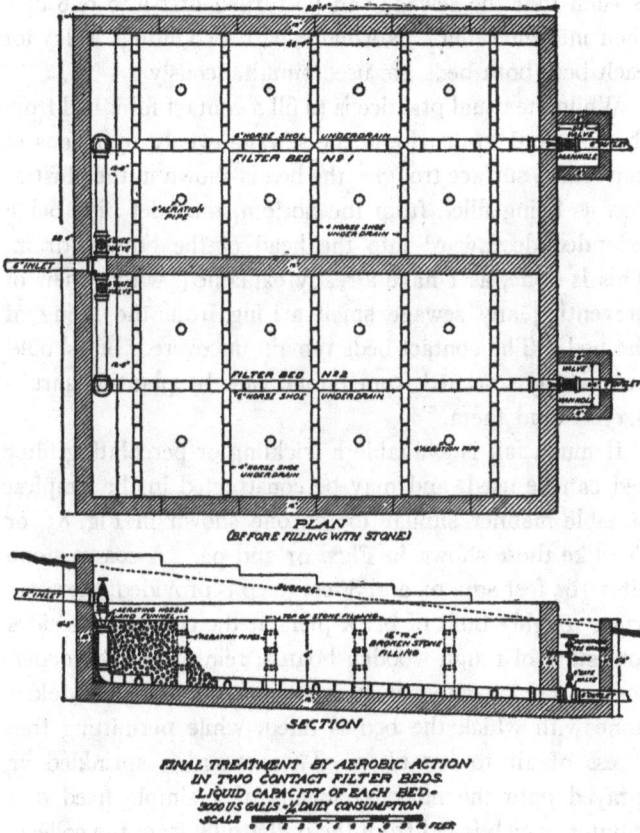

FIG. 88. — PLAN AND SECTION OF CONTACT FILTER BEDS FOR
DISPOSAL OF SEWAGE FROM A LARGE BUILDING.

closed by means of a gate valve operated by the attendant. At the inlet to the contact bed the arrangement of the piping is such that the sewage can be turned first into one bed, then into the other. Normally, with two fillings a day for each bed, both beds are used simultaneously.

While the usual practice is to fill a contact filter bed from the top, and to distribute the sewage evenly by means of perforated surface troughs, the bed is shown in the illustration as being filled from the bottom, the inlet pipe being extended downward into the head of the bottom drain. This is done, as I have already explained, with a view of preventing any sewage smell arising from the filling of the bed. The contact beds remain uncovered, but should be suitably screened from public view by planting shrubbery around them.

If much fall is available a trickling or percolating filter bed can be used, and may be constructed in the simplest possible manner similar to the one shown in Fig. 85, or else like those shown in Figs. 91 and 92. A coarse stone filter $16\frac{1}{2}$ feet square, and 8 feet deep is provided, the same being roughly built of brick piers at the corners, the sides consisting of rough wooden boards, reinforced by wooden posts, all put together in such a way as to hold the broken stone with which the bed is filled, while permitting free access of air to the filter. The sewage is sprinkled or sprayed onto the filter by means of a simple fixed distributor, which is fed from the outlet pipe from the collecting tank, the flow being controlled by a gate valve so as to accomplish the proper slow dosing of the filter. At the bottom of the filter a layer of concrete is provided, and the bed is suitably underdrained. The purified sewage is discharged by means of an outlet drain, which has no gate valve, into a trench filled with broken stone, and emptying on the slope to the lake.

VIII. In Fig. 89 is shown the view of a large septic tank in combination with four contact filter beds, installed for the purification of the sewage from the summer hotel illustrated in view in Fig. 35, for an average population of 700 people. Figure 90 is a plan of this disposal

FIG. 89. — VIEW OF CONTACT FILTER BEDS, DOSING CHAMBERS AND SEPTIC TANK FOR A LARGE HOTEL.

plant, which was designed and its execution superintended by the writer.

The photographic view shows the top of the septic tank, which is arched over and provided with four cleaning manholes. Between the septic tank and the contact beds are located the collecting tanks, two in number, as the

PLAN OF SEPTIC TANK & CONTACT FILTER BEDS

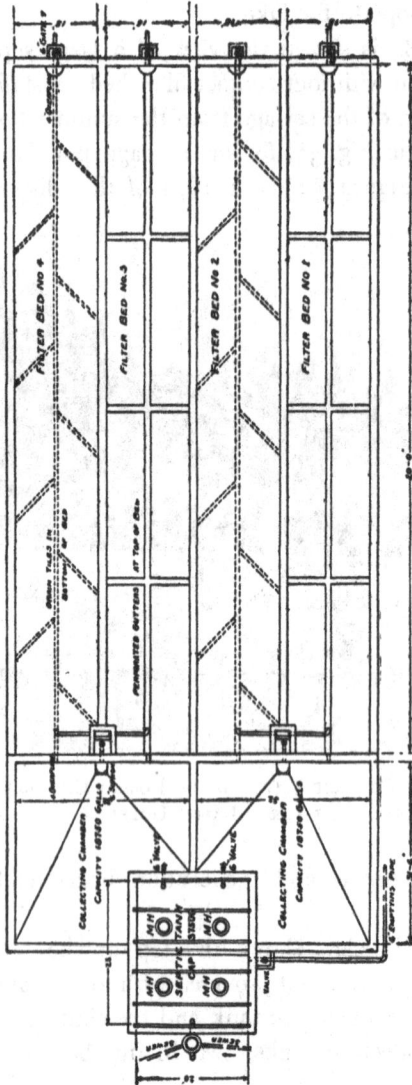

LONGITUDINAL SECTION

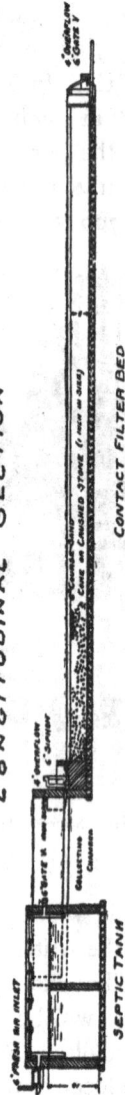

FIG. 90. — PLAN AND SECTION OF SEPTIC TANK AND CONTACT FILTER BEDS FOR A LARGE HOTEL.

plan shows. The filter beds are shown without any filling material, which was put into them after the view was taken.

The septic tank is 20 feet 6 inches by 25 feet, and 10 feet deep. It has a capacity of 37,500 gallons or half the daily amount of the sewage flow, which is estimated at 75,000, assuming 100 gallons of sewage as the maximum for each person, the number of people contributing being taken at 750. The septic tank is divided into two parts by a cross wall, over which the sewage flows when going from the first to the second chamber. Both inlet and outlet pipes are submerged.

There are two collecting chambers for the liquid or septic sewage. Each chamber is 31 feet 4 inches by 31 feet 6 inches, and has an average depth of 3 feet. The capacity of each collecting tank is therefore one half of the capacity of the septic tank, or 18,750 gallons. By means of gate valves sewage can be made to flow into either of the two chambers. These chambers are left open to the air.

Each chamber is provided with an automatic sewage siphon which empties the chamber as soon as it has become filled. The two siphons discharge into a well or siphon chamber, provided with two distributing shear gates. In this way sewage from collecting chamber I can be discharged over filter beds No. 1 or 2, and sewage from chamber II can be discharged into filter beds No. 3 and 4 alternately.

The four contact filter beds are each of a liquid capacity of 18,750 gallons, corresponding to the discharge of the siphon. Each filter bed is 120 feet long and 15 feet wide, and has an average depth of 3 feet. The bottom of each bed is underdrained by a number of lines of horseshoe drain tiles, 4 inches and 6 inches in diameter. The sewage is distributed on top of the bed by means of perforated troughs or gutters, running both longitudinally and crosswise over the beds. This method of distribution did not prove altogether a success, because the metal troughs were apt to warp.

The beds are filled with various materials. Coke and broken stones were tried. The size of these was 1 inch. On top a layer of coarse sand was put, which afterward proved to be a mistake.

At the lower end of each filter bed there is a manhole, in which the gate valve is set which controls the filling and emptying of the bed. The outlets are 6 inches in diameter. Each bed is also provided with a 6-inch overflow pipe in case the opening of the gate valve should be forgotten at the proper time by the attendant. Each bed is intended to stand filled with sewage for the period of 2 hours. The discharge from the four filter beds continues to a large manhole, from which samples of the effluent may be taken, and it finally empties into a river below the normal water level.

IX. *Trickling Sewage Filters.*

Figs. 91 and 92 illustrate in view and in section the arrangement of trickling filters which have been used for the disposal of large volumes of sewage. The dosing of the filter is effected by means of revolving sprinkler arms.

The dotted double lines in the center of the filter, Fig. 92, indicate drain pipes placed in the filter for its aëration. The view, Fig. 91, shows the sprinkler in action. Note particularly that the filter stands entirely free, being built up on a bed of concrete, which is provided with

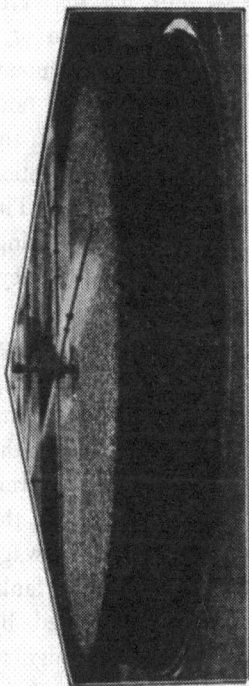

FIG. 91.—VIEW OF LARGE TRICKLING FILTER WITH SPRINKLER IN OPERATION.

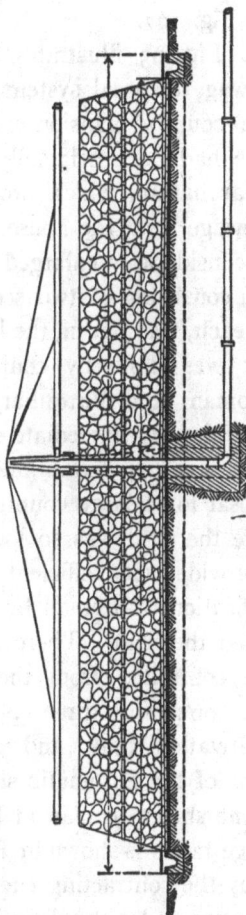

FIG. 92.—SECTION THROUGH A LARGE TRICKLING FILTER.

drainage. Such well-aërated filters give a better effect than those which are built into the ground as in illustration, Fig. 105.

X. Fig. 93 illustrates a topographical plan, showing a sewage disposal system by subsurface irrigation for a large country mansion at Locust Valley, L. I. This house had been originally provided with a system of a similar nature, which proved too small for the amount of sewage from the house. Accordingly the sewage tank was considerably enlarged, and a new field, shown in the plan, consisting of two sections of 10 lines of absorption tiles each, was put in, the lines being laid as shown. The work was done by contract by the firm of Waring, Chapman, and Farquhar, from my plans and surveys and under my immediate direction.

XI. Fig. 94 shows the plan of a system of sewage disposal for a large country house on the Hudson River, where the field was so located that it became necessary to provide very efficient underdrainage, and also the rectification of a small brook which during heavy storms flooded the field. There are two sections of tiles in the field, controlled from the gate chamber. The sewage tank, comprising three compartments, viz., a septic tank, a cultivation tank, and a liquid chamber emptied by means of the automatic siphon, was similar in design to the one shown in plan in Fig. 80. A cross-section of the sewage tanks is shown in Fig. 95. The work was carried out by the contracting engineers, Waring, Chapman, and Farquhar under the direction of the author.

For the stable and barn belonging to the country house described in the preceding example quite an elaborate sewage disposal system was installed, comprising a septic

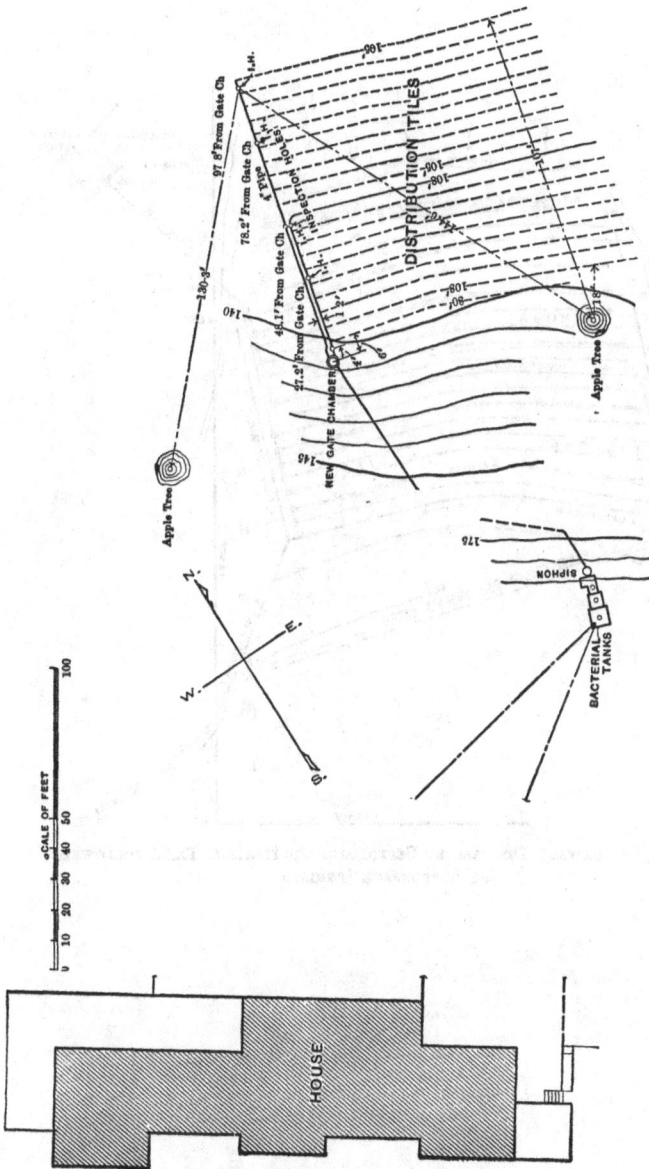

FIG. 93. — SEWAGE DISPOSAL BY SUBSURFACE IRRIGATION.

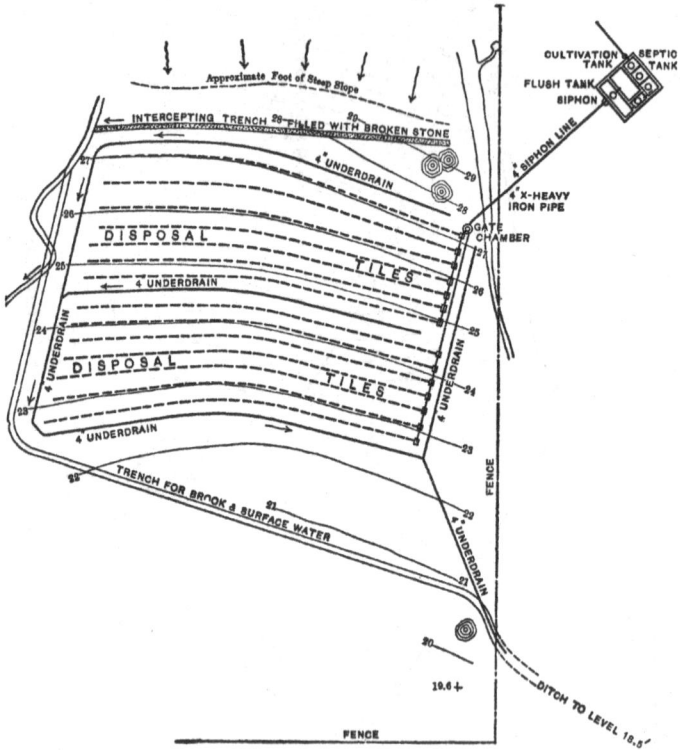

FIG. 94. — SEWAGE DISPOSAL BY SEPTIC AND CULTIVATION TANK, FOLLOWED BY SUBSURFACE IRRIGATION.

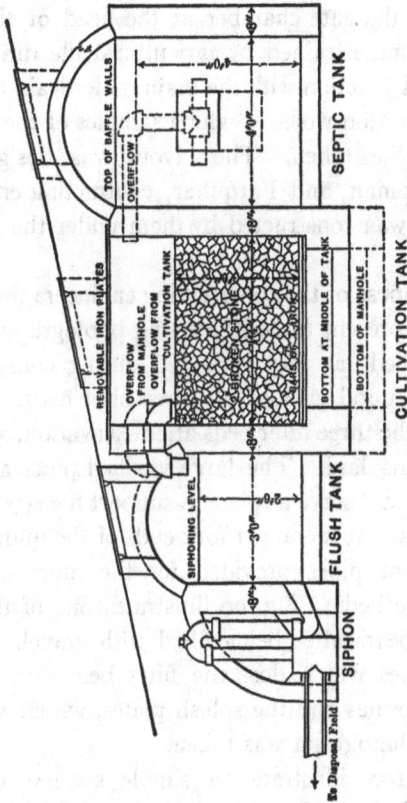

Fig. 95. — Cross-Section through the Sewage Tanks.

tank with dosing chamber and siphon, and three intermittent gravel filter beds, arranged as shown in the plan,
Fig. 96. Two sections through the three filter beds are
also shown. The sewage is turned onto these alternately
by means of the gate chamber at the head of the beds.
Each bed is underdrained by agricultural tile drains, and
at the point of junction with the main underdrain there are
provided inspection wells, at which samples of the purified
sewage may be taken. The layout was designed by
Waring, Chapman, and Farquhar, contracting engineers,
and the work was constructed by them under the author's
direction.

By the kindness of the constructing engineers the author
is enabled to present three interesting photographic views
of the filter beds as they appeared during construction.
Fig. 97 is a general view of all three filter beds. Fig. 98
shows one of the three filter beds after excavation, and with
the underdrains laid. The large vitrified pipes are filled
with concrete and serve as piers to support the sewage distribution pipes. At the upper four ends of the underdrains
are shown vent pipes provided for the more complete
aëration of the beds. Fig. 99 illustrates one of the filter
beds as it appears after being filled with gravel. It also
shows the pipes which dose the filter bed with sewage,
without the nozzles and the splash plates, which were put
on after the photograph was taken.

XII. Fig. 100 illustrates a simple sewage disposal
plant installed by the New York Sewage Disposal Company at Mount Kisco, N. Y. The plan shows the location
of the residence, of the sewage tank, and of the sewage
field, which latter is laid out in two equal sections, each
having five rows of irrigation tiles.

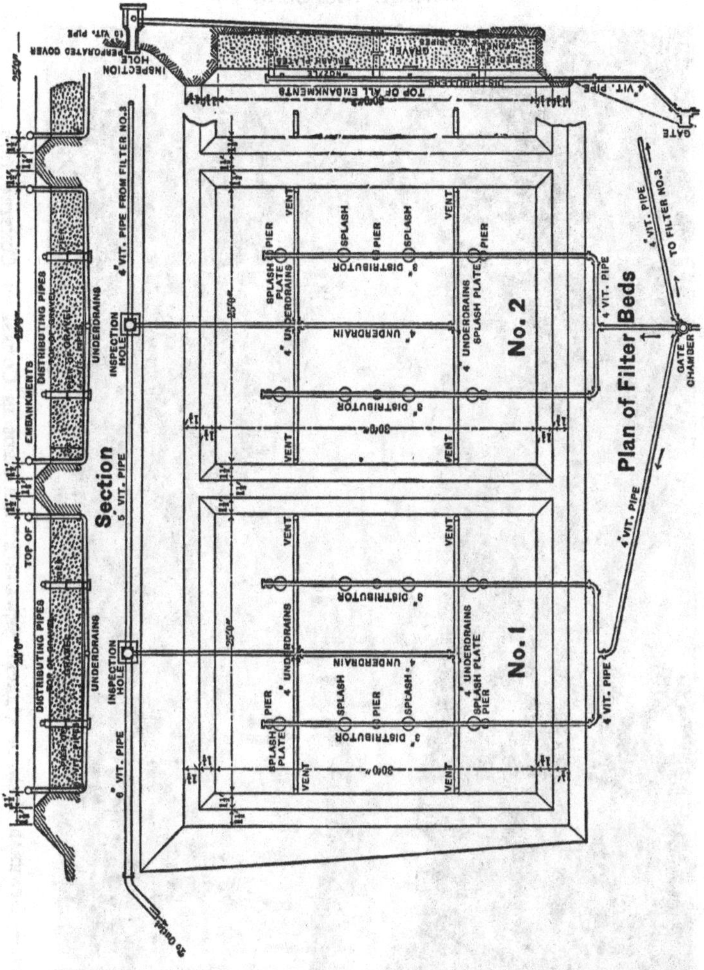

FIG. 96. — SEWAGE DISPOSAL BY GRAVEL FILTRATION BEDS.

Fig. 97. — General View of Gravel Sewage Filter Beds in Course of Construction.

FIG. 98. — VIEW OF ONE FILTRATION BED, SHOWING CONSTRUCTION OF UNDERDRAINS.

Fig. 99. — View of one Filtration Bed, filled with Gravel, with Sewage Distribution Pipes at Top of Bed.

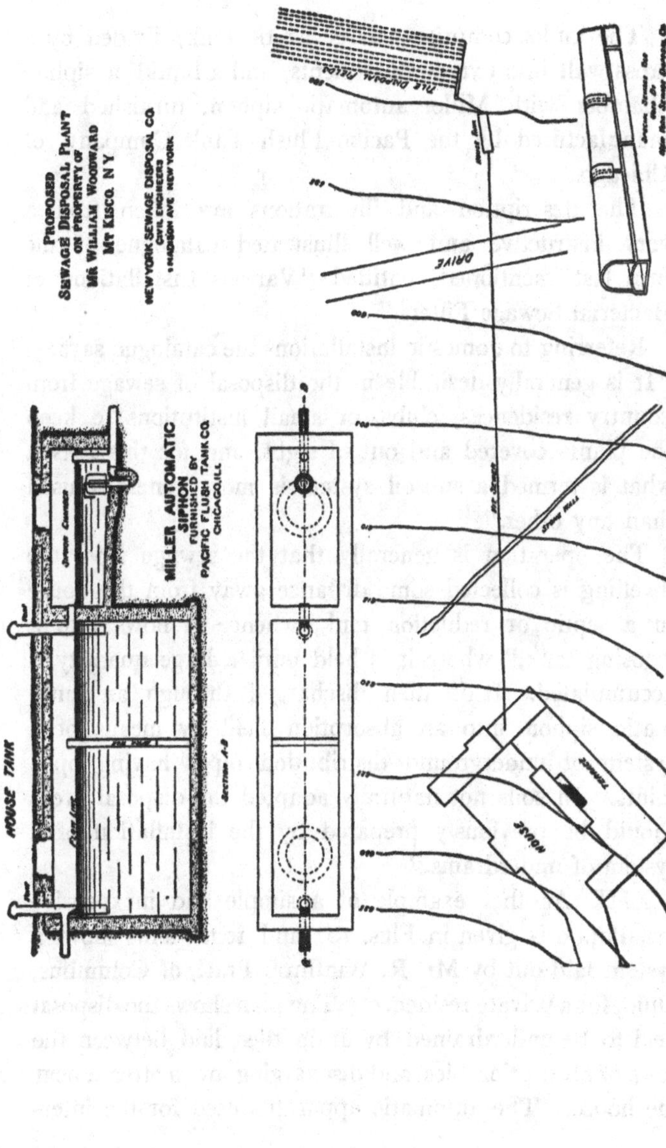

FIG. 100. — SEWAGE DISPOSAL BY SUBSURFACE IRRIGATION.

The tanks comprise a large septic tank, divided by a cross wall into two compartments, and a liquid or siphon chamber with Miller automatic siphon, furnished and manufactured by the Pacific Flush Tank Company, of Chicago.

The description and illustrations are taken from a very instructive and well illustrated catalogue of the firm last mentioned, entitled "Various Installations of Bacterial Sewage Filters."

Referring to domestic installations the catalogue says:— "It is generally desirable in the disposal of sewage from country residences, clubs, or small institutions to keep the plants covered and out of sight, and for this reason what is termed a subsoil system is more generally used than any other."

The operation is generally that the sewage from the dwelling is collected some distance away from the house in a septic or reduction tank, whence it flows into a "dosing tank," where it is held until a large quantity is accumulated. It is then discharged through an automatic siphon into an absorption field by means of a system of underground distribution pipes having open joints. In soils not naturally adapted the disposal areas should be previously prepared by the installation of a system of underdrains."

XIII. Another example of a simple and inexpensive installation is given in Figs. 101 and 102. This shows a system laid out by Mr. R. Winthrop Pratt, of Columbus, Ohio, for a private residence. The plan shows the disposal field to be underdrained by drain tiles, laid between the rows of absorption tiles, and discharging into a stream near the house. The automatic apparatus used for the inter-

mittent discharge of the sewage from the dosing tank is
the siphon made by the Pacific Flush Tank Company of
Chicago. Similar apparatus is made by the Newport,
R. I., Foundry Company; and the well-known engineers,

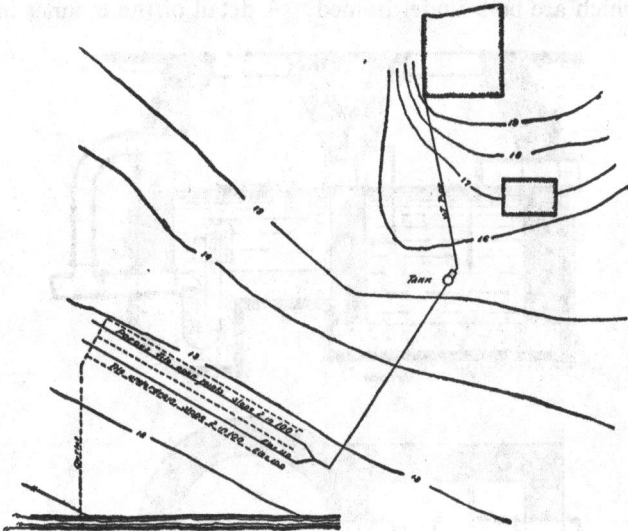

SEWAGE DISPOSAL
FOR RESIDENCE OF
O. T. CORSON, ESQ.
NEAR COLUMBUS, OHIO
FEB 1905.   R WINTHROP PRATT, ENGR.
COLUMBUS, OHIO

FIG. 101. — SEWAGE DISPOSAL BY SUBSURFACE IRRIGATION.

Waring, Chapman, and Farquhar, manufacture another
siphon of their own design.

XIV. Fig. 103 illustrates the sewage disposal system
designed by Mr. Wm. S. McHarg, of Chicago, for a
number of farm buildings of an estate.

The illustration shows the design for the sewage tank, which contains three chambers with baffle walls; the dosing chamber contains the Miller and Adams automatic siphon. The sewage is discharged into two disposal fields, which are both underdrained. A detail of the manner in

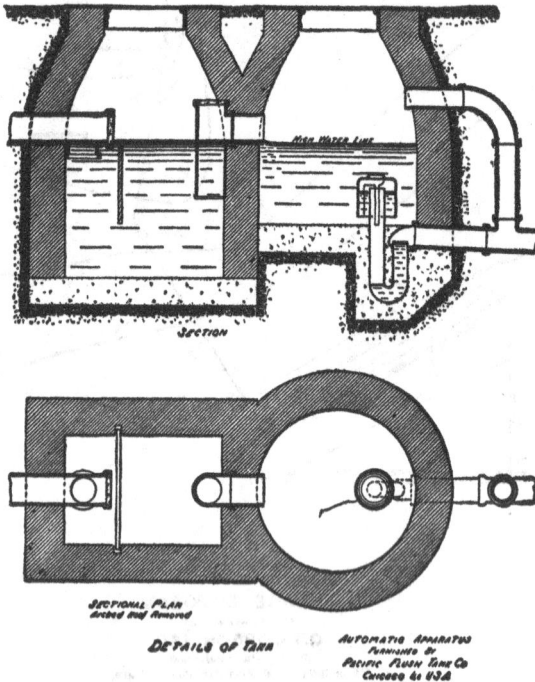

FIG. 102. — DETAILS OF SEWAGE TANK.

which the disposal tiles have been arranged is also given. The illustration is taken from the above-mentioned catalogue of the Pacific Flush Tank Company.

XV. The example of a contact filter bed system shown in Fig. 104 illustrates the sewage disposal plant for

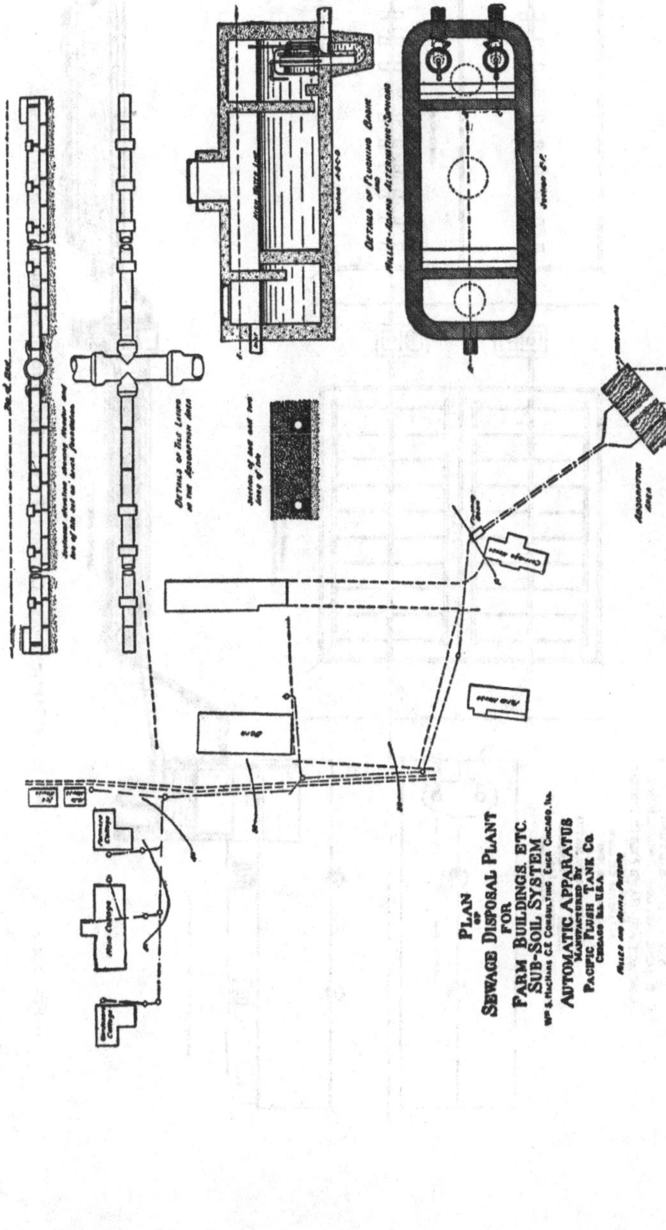

FIG. 103. — SEWAGE DISPOSAL BY SUBSURFACE IRRIGATION.

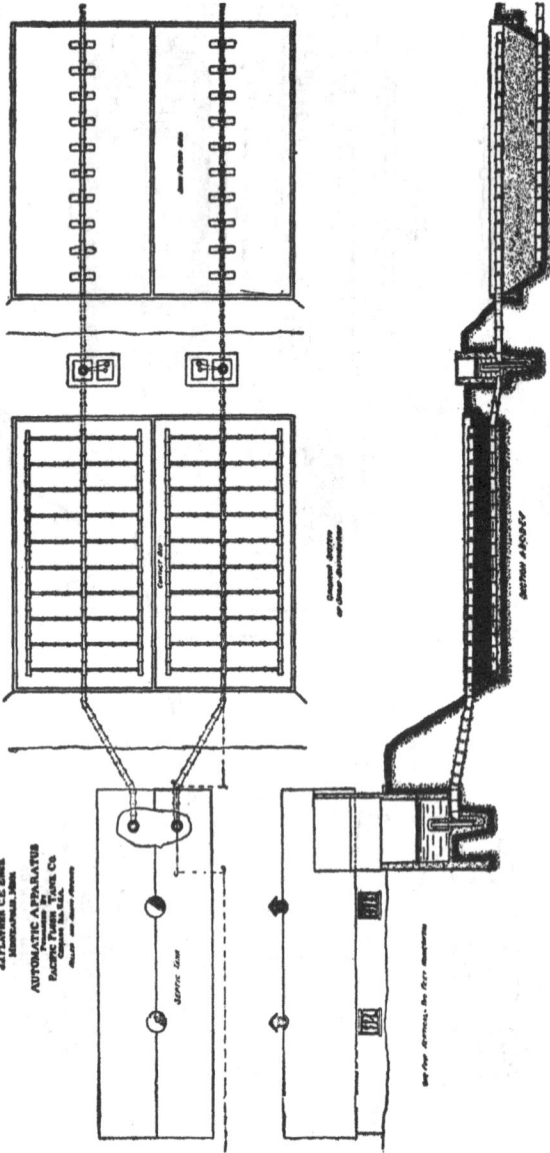

PLAN
SEWAGE DISPOSAL PLANT
INSANE ASYLUM, ANOKA, MINN.
CONTACT AND SAND FILTER SYSTEM
FOUR BEDS
All Plans by C.E. Eng.
Minneapolis, Minn.

AUTOMATIC APPARATUS
Furnished By
PACIFIC FLUSH TANK CO.

FIG. 104. — SEWAGE DISPOSAL BY SEPTIC TANK, CONTACT FILTER BED AND SUBSEQUENT SAND FILTRATION.

a Western insane hospital, designed by J. L. Flather, C.E., of Minneapolis, Minn.

"This system of sewage purification," says the catalogue, "requires much less depth and area than intermittent sand filtration and is found of value when either depth or area is insufficient.

"In operating plants of this type, the sewage flows from a septic or reduction tank into one or more distributing chambers, thence through air lock feeds with very little loss of head into the beds in rotation, where it is held in contact with the filtering material, which should be quite coarse, for a given length of time, and is then discharged through "timed siphons," one being located in each bed.

"The Adams air lock feeds are installed in the distributing chamber, one for each bed, and are intended to feed the sewage into the contact beds in rotation, filling each, cutoffs being used to skip any one or more of the beds.

"In each bed is located a Miller 'timed siphon,' which holds the sewage in contact with the bed for a given length of time and then completely drains the bed."

The example shows two septic tanks, and two dosing chambers with Miller-Adams double alternating siphons, which discharge the sewage on to two contact beds which are called primary beds. The Miller timed siphons empty these beds and throw the sewage on to a set of two secondary contact beds, having a finer filtering material. The details of these automatic appliances may be looked up in the catalogue of the Pacific Flush Tank Company, from which this example is taken.

XVI. In Fig. 105 I illustrate a sewage disposal system by means of a septic tank and a percolating filter. This was designed and executed by Mr. H. M. Reel, C.E., of Youngstown, Ohio.

The upper part of the cut shows the septic tank and the percolating filter in section, the lower part shows the plan of both tanks. To the right is an illustration of the Miller automatic siphon used to dose the percolating filter automatically at intervals.

The several baffle boards in the septic tank are designed to precipitate the solid matters and to prevent their being carried over by the siphon.

The size of the dosing tank is quite small. The percolating or trickling filter is shown to be aërated on the four sides, and to be filled with washed and screened gravel. The dosing is accomplished by means of a network of pipes with a number of sprinkling orifices or nozzles. The trickling filter is covered over to prevent the spread of sewage odors.

Regarding percolating filters the catalogue of the Pacific Flush Tank Company speaks as follows: — "On account of the rapid rate of filtration and small area required as compared with sand filters and also contact beds, percolating filters have recently become quite popular in our country and abroad.

"A much greater depth is needed to install this type of sewage disposal. . . . Filters of this type are necessarily of coarse material and the distribution of the sewage so as to cover the whole surface of the filter becomes an important feature. The installation shown in this example is by means of a system of fixed spray distributors. . . . This type of sewage purification does not give as high a

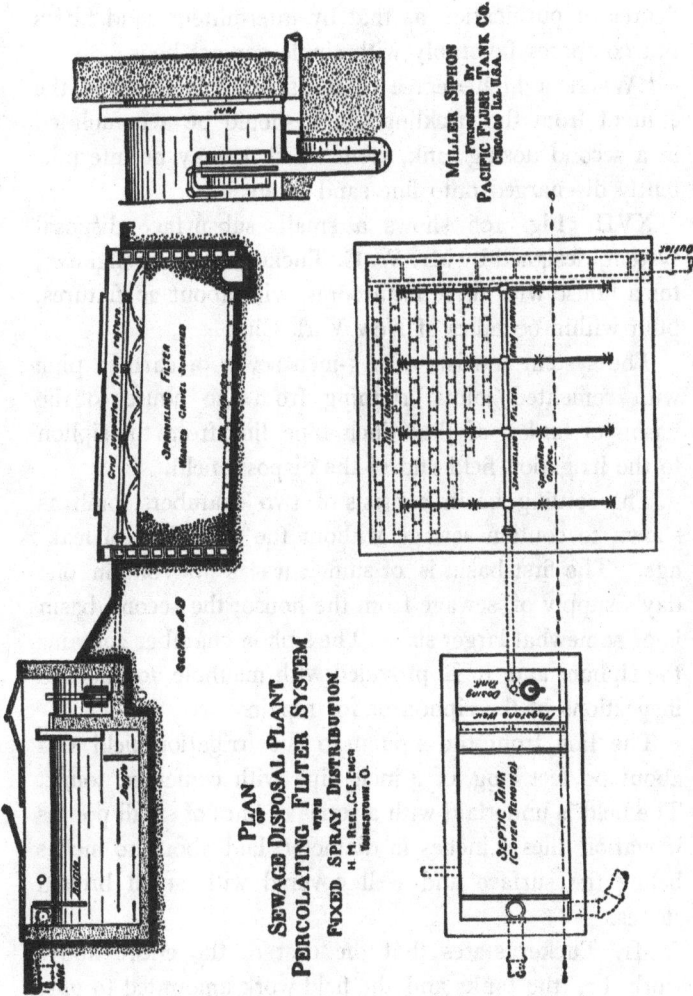

FIG. 105. — SEWAGE DISPOSAL BY SEPTIC TANK AND SPRINKLING FILTER.

degree of purification as that by intermittent sand filters but compares favorably with single contact beds.

"Where a high degree of purification is required the effluent from the trickling filters should be accumulated in a second dosing tank, from which it may be intermittently discharged onto fine sand filters."

XVII. Fig. 106 shows a small subsurface disposal system, designed by Mr. W. C. Tucker, sanitary engineer, for a house with three bathrooms, with about 20 fixtures, built within 60 miles of New York City.

The system consists of a 5-inch sewer of earthen pipe with cemented joints, running from the house to the basins or tanks; of the 4-inch pipe line from the siphon to the irrigation field and of the disposal field.

The settling basin consists of two chambers, built as shown to contain sewage without the possibility of leakage. The first basin is of sufficient size to contain one day's supply of sewage from the house; the second basin is of somewhat larger size. The siphon chamber contains the siphon, and it is provided with manhole for needed inspections of the siphon or for repairs.

The line from the siphon to the irrigation field was about 60 feet long, of 4-inch pipe with cemented joints. The field is underlaid with a series of lines of small porous irrigation tiles 3 inches in diameter, laid about 10 inches below the surface and well covered with small broken stones.

Mr. Tucker states that the cost of the entire above work, i.e., the tanks and the field work amounted to only $230. This is more than exceptionally low, and the author does not know of another instance where such work was constructed so cheaply.

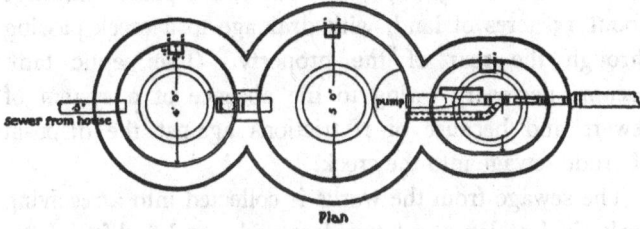

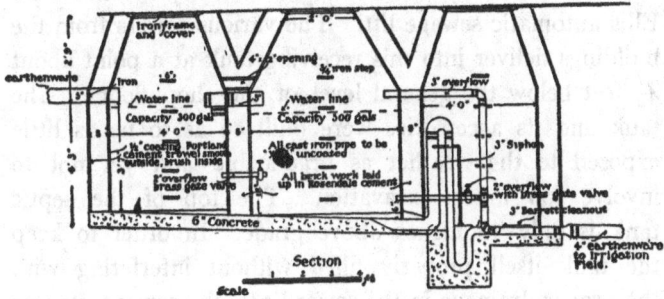

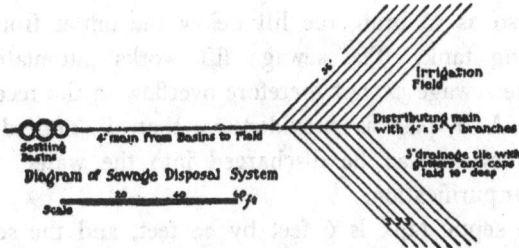

FIG. 106. — A SMALL ECONOMICALLY BUILT SUBSURFACE
IRRIGATION SYSTEM.

XVIII. A septic tank for the purification of the sewage from a manufacturing plant located at Aldene, N. J., is illustrated in Fig. 107. The site of the plant comprises about 15 acres of land, with drainage to a creek passing through the rear of the property. The septic tank became necessary owing to the absence of a system of sewers and because of restrictions against the disposal of crude sewage into the creek.

The sewage from the works is collected into a receiving tank six feet long and two feet wide, and is lifted from there automatically into the septic tank by means of an Ellis automatic sewage lift. The various sewers from the buildings deliver into this receiving tank at a point about 4½ feet below the general level of the shop floors. The tank and its accessories were built so as to be as little exposed to the weather as practicable and yet not to involve too much excavation. The top of the septic tank is only 18 inches above grade. In order to keep the tank itself properly filled without interfering with the proper drainage in the sewer lines, the sewage lift was used, and it is located, as shown, in a pit about 10½ feet deep, so as to bring the lift below the outlet from the receiving tank. The sewage lift works automatically, and the sewage cannot therefore overflow in the receiving tank. A by-pass is provided so that, if desired, the crude sewage can be discharged into the water course without purification.

The septic tank is 6 feet by 20 feet, and the sewage is held at a depth of 6 feet 3 inches. Its size is so designed that the sewage will pass through the tank at a very slow velocity. The outlet drain of the septic tank is a 6-inch pipe submerged about three feet in the tank so

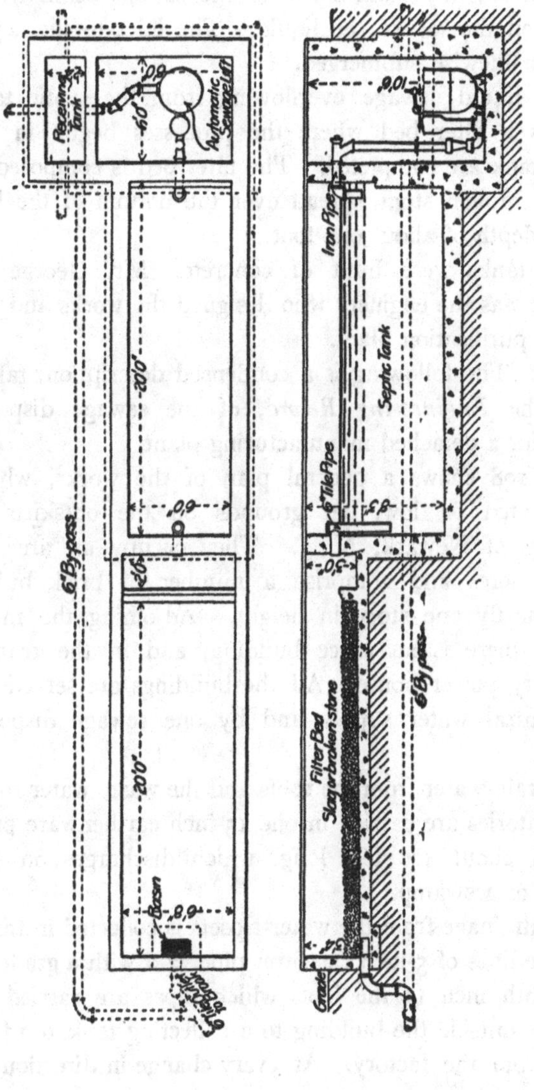

FIG. 107.—SEWAGE DISPOSAL BY SEWAGE LIFT, SEPTIC TANK AND FILTRATION BED.

as to avoid any disturbance of the sewage scum which forms on the top of the liquid. The inlet to the septic tank is likewise submerged.

The liquid sewage overflowing from the septic tank reaches a filter bed where the processes begun in the septic tank are completed. The filter bed is composed of slag or broken stone spread over the bottom of the bed to the depth of about one foot.

The tanks were built of concrete. Mr. George K. Hooper was the engineer who designed the works and the sewage purification plant.

XIX. The following is a condensed description, taken from the *Engineering Record*, of the sewage disposal works for a detached manufacturing plant.

Fig. 108 shows a general plan of the works, which are located on low, flat grounds on the outskirts of the city of Newark, N. J. They occupy an area of several acres and comprise a number of brick buildings, mostly one story in height. Adjoining the main factory there is an office building, and in the rear a two-story power house. All the buildings are served by one central water supply and by one sewage disposal system.

The rain-water from the roofs and the waste water from the lavatories are carried in one 15-inch earthenware pipe channel about 400 feet long, which discharges on the surface of a swamp.

The drainage from the water-closets is collected in three separate lines of 5-inch cast iron pipe, laid with a grade of one-eighth inch to the foot, which pipes are carried in trenches outside the building to a collecting tank 100 feet away from the factory. At every change in direction in

the sewer line there is a cleanout made accessible by a
brick pit or vault, with cast-iron cover.

The tank is 6½ feet by 7 feet on the inside, and 15 feet
deep. The bottom is of concrete, 18 inches thick, the

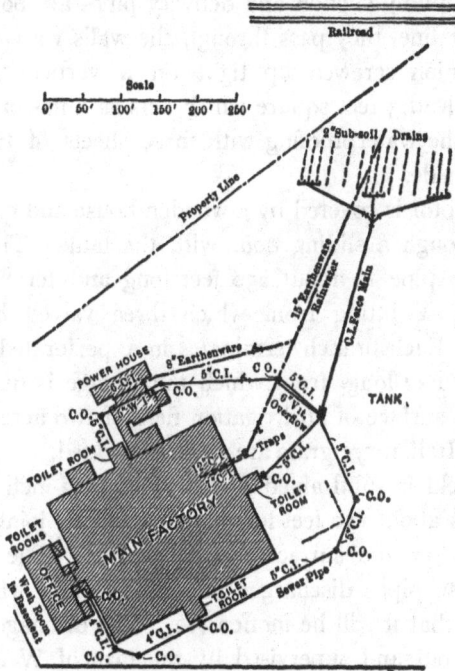

PLAN OF WORKS AND SEWERS

FIG. 108. — SEWAGE DISPOSAL BY SUBSURFACE IRRIGATION FOR A
MANUFACTURING PLANT.

walls are of brick, and the top of reinforced concrete
4 inches thick.

At one side of the tank is the 2½ by 6½ foot chamber,
9 feet deep, in which there is a 2 horse-power electric
motor driving a 3-inch centrifugal pump. The pumping

is started and stopped automatically. The walls of this tank are of brick, and into them and the concrete floor is built a waterproof course consisting of six thicknesses of tar paper laid in hot asphalt.

As the pump suction and delivery pipes are both below the water line, they pass through the walls with a double flange union screwed up tight on a vertical sheet of 5-pound lead 3 feet square which acts as a flashing and is built in the waterproofing with three sheets of tar paper on each side.

The motor is covered by a wooden house and communicates through a sliding door with the tank. The pump discharge pipe is about 250 feet long and terminates in a double Y fitting from which three valved branches radiate. Each branch terminates in a perforated header about 50 feet long, from which the sewage is discharged on to the surface of an irrigation field of two acres, where crops of Italian rye grass are to be cultivated.

The field is subdrained by ten lines of 2-inch agricultural tiles about 150 feet long, laid with open joints, about 4 feet below the surface of the ground. The effluent from these pipes discharges into the swamp, where it is expected that it will be inoffensive. The drainage system was laid out and supervised by the firm of Walker and Stidham, engineers.

XX. The sewage from the house, illustrated in Fig. 68 (Part II), estimated at a maximum daily flow of 1000 gallons, is conveyed through a 6-inch earthenware sewer pipe, about 600 feet long, to a series of settling basins, located about 50 feet below the level of the house. Fig. 109 shows the sewage settling basins in plan and in cross-section. The main sewer pipe is laid with a uniform

grade and with a straight alignment between the manholes
On this line of house sewer there are provided four brick
manholes at points where it is convenient to change the
direction and the grade. The settling basins are circular,
with brick walls and concrete bottom and with iron cor-

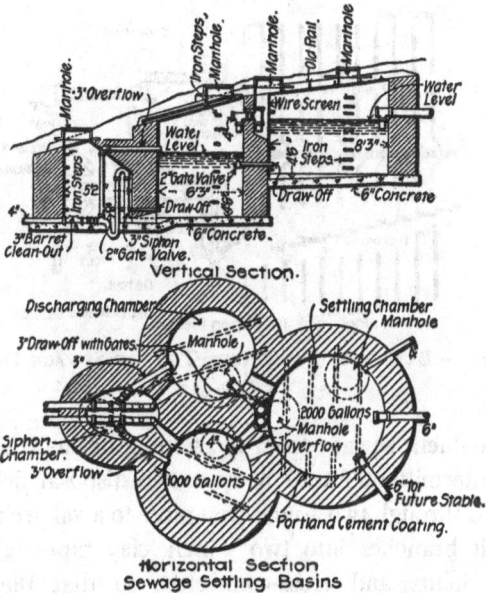

FIG. 109. — SEWAGE TANKS FOR HOUSE.

poration manhole frames and covers, supported on iron
rail beams. They are built on a side hill, as shown in
section, and their bottoms are offset to conform to the
slope of the surface. The water overflows freely from
the highest settling basin, where it is received into the
twin siphon chambers and accumulates there until it
fills them to the level required to operate the discharging

siphons, which automatically deliver the sewage to the respective irrigation fields. Each of the basins is provided with an emptying pipe by which most of the liquid sewage can be drawn off, and has manholes through which the accumulated solid matter can be removed.

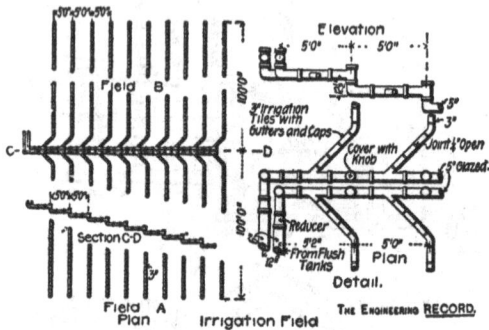

FIG. 110. — LAYOUT OF SUBSURFACE TILES IN SEWAGE DISPOSAL FIELD.

The effluent sewage from the settling basins is disposed of by intermittent filtration upon two disposal fields. It is carried through two 4-inch iron pipes to a valve chamber, where it branches into two 5-inch clay pipes laid with cement joints and cross-connected, so that the whole discharge from both flush tanks or the discharge from either flush tank may be diverted to either field. The valve chamber allows the sewage to be alternated between the fields so that they can be rested and dried out whenever desired. These pipes are connected to the distribution system, which disposes of the sewage at a depth of about 10 inches below the surface of the soil. As the irrigation fields have a surface slope of about 1 in 5, the main delivery pipes are laid with special connections making a drop

every 5 feet (see Fig. 110), and between these points the
pipe is level. At each drop a covered handhole is provided to give access to the interior of the pipe for inspection and cleaning. The lateral branches are taken out

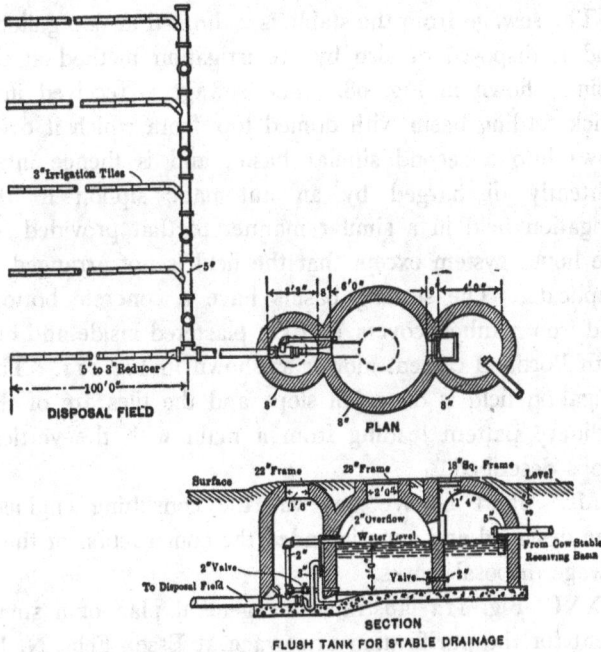

FIG. 111. — SEWAGE TANK AND DISPOSAL FIELD FOR STABLE.

though Y's between these points. All branches are
about 100 feet long, are open at the outer end and are
composed of 3-inch irrigation tile laid with gutters and
caps, made with open joints and surrounded with broken
stones. The entire disposal field is underdrained to a
depth of 5 feet.

The road drainage, the rainfall runoff and the discharge from the roof leaders, are all received and conveyed by a system of catch basins and clay pipes from 3 to 6 inches in diameter, laid with cemented joints, and disposed of by surface filtration at various points.

The sewage from the stable is estimated at 600 gallons, and is disposed of also by the irrigation method at the points shown in Fig. 68. The sewage is received in a brick settling basin with domed top, from which it overflows into a second similar basin, and is thence intermittently discharged by an automatic siphon to the irrigation field in a similar manner to that provided for the house system except that the field is not arranged in duplicate. The settling basins have a concrete bottom and iron manhole covers, and are plastered inside and out with Portland cement mortar as shown in Fig. 111. The irrigation field is on a hill slope and the tiles are of the ordinary pattern leading from a main with the vertical drops described.

Mr. Albert L. Webster was the consulting engineer who designed and superintended the construction of these sewage disposal works.

XXI. Fig. 112 illustrates the general plan of a small plant for the purification of sewage at Essex Fells, N. J., and is an interesting and useful example of a satisfactory and economical method of treatment. Essex Fells is a picturesque residential district in New Jersey, populated chiefly by business men from New York.

Originally this settlement of houses had a small disposal plant designed with the best practice obtaining at the date of installation. Briefly described, the plant consisted of a circular grit chamber 18 feet in diameter

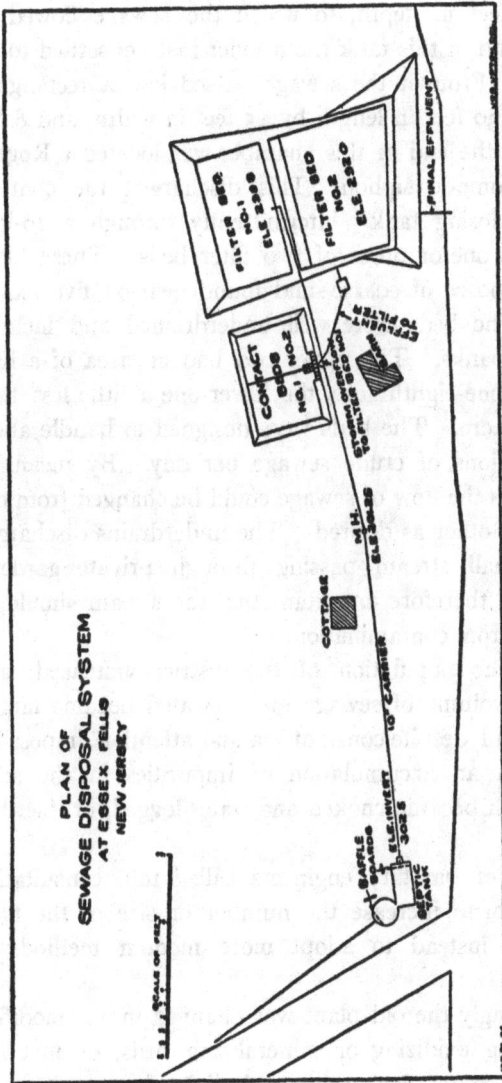

PLAN OF
SEWAGE DISPOSAL SYSTEM
AT ESSEX FELLS
NEW JERSEY

SCALE OF FEET

FILTER BED NO. 1
ELE 238.5

FILTER BED NO. 2
ELE 236.0

CONTACT BEDS NO. 1 NO. 2

FINAL EFFLUENT

EFFLUENT TO FILTER

STONE CHECKDAM TO FILTER BED

M.H. ELE 248.55

TO FARMER

COTTAGE

ELE INLET 302.5

BAFFLE BOARD

SEPTIC TANK

Fig. 112.—Sewage Disposal by Septic Tank and Contact Filter Beds with "Air-Lock" Automatic Siphons.

and 11 feet in depth, to which the sewage flowed by
gravity, and in this tank the heavier matters settled to the
bottom.   From it the sewage passed into a rectangular
chamber, 30 feet in length by 15 feet in width, and 8 feet
deep.   At the end of this chamber was located a Rogers-
Field automatic siphon.   This discharged the contents
of the "dosing tank" intermittently through a 10-inch
pipe on to one or other of two filter beds.   These filters
were composed of coarse sand found near-by, five feet in
depth.   The beds were well underdrained and inclosed
by earth banks.   The upper bed had an area of a little
less than one-eighth acre, the lower one a little less than
one-sixth acre.   The beds were designed to handle about
20,000 gallons of crude sewage per day.   By means of
sluice gates the flow of sewage could be changed from one
bed to the other as desired.   The underdrains discharged
into a small stream passing through private gardens,
and it was therefore important that the stream should be
kept free from contamination.

When the population of the district increased, and
the daily volume of sewage to be treated became larger,
it was found, despite constant manual attention, impossible
to prevent an accumulation of impurities in the sand
beds, which became choked and water-logged, or "sewage
sick."

A firm of sanitary engineers called into consultation
decided not to increase the number or size of the filter
beds, but instead to adopt more modern methods of
disposal.

Accordingly the old plant was changed into a modified
one, having oxidizing or mineralizing beds, or in other
words, the contact bed system.   A slight change was made

in the old dosing tank, and two contact beds were constructed, each 35 feet by 50 feet, and 3 feet in depth. A deep baffle board was placed across the inlet to the grit chamber and another across the outlet end of the rectangular tank with a weir, over which the sewage falls instead of being emptied by a siphon. In this way the dosing tank was changed into a "septic tank." A 10-inch carrier conveys the septic sewage to a distributing chamber constructed at the entrance to the two contact beds. Here a very ingenious device, known as the "air-lock" method is employed. It is automatic and has no moving parts, but differs from a siphon by not requiring any fall. By this device each of the beds is filled alternately.

The principle upon which this method of automatic operation is based was first worked out in England by Mr. Adams, C.E., but it was modified in this country by Mr. Albert Priestman, of Philadelphia, Pa., who designed the above sewage disposal plant, and from whose description I have condensed my own. In the conclusion of his very readable article in "House and Garden," he states "the use of the present combination of processes requires less area than the former method, while it is capable of taking care of the continually increasing volume of sewage for many years to come. It is also more automatic in operation, seeing that it is not necessary to manipulate any valves by hand, except at infrequent intervals."

XXII. Some years ago a small sewage disposal plant was built at the Eastern Indiana Hospital for the Insane, which contained then a population of about 800 persons, and used an average daily amount of water estimated at 80,000 United States gallons.

Formerly the sewage was discharged directly into a

stream about 1500 feet away, but the stream had a very small dry-weather flow, and the presence of the sewage became quite noticeable for a distance of two miles down stream. A septic tank was finally adopted for overcoming the nuisance, and the system has been described by Prof. R. L. Sackett in a paper read before the Indiana Engineers' Society.

The sewage enters the septic tank through a goose neck, and is discharged into a grit chamber, through which it flows into the septic tank proper. Here it stands at a depth of from 6 to 8 feet, and for a period which seems to secure the best action of the bacteria. The structure, which is 67 feet long, 20 feet wide, and 10 feet deep, is built of Portland cement concrete. In order to render the walls waterproof a coating of cement mortar was troweled over the interior surface exposed to the water. Various valves are provided, by means of which the chambers are cleaned and the sludge discharged on the ground. The tank is housed in and has a slate roof.

From the tank the sewage is led into a so-called dosing chamber, which contains special apparatus, made by W. S. Shields, C.E., of Chicago, to discharge a quantity of the sewage automatically through pipe lines on to four filter beds. Each of the filter beds is 100 feet square. To each bed there are three cypress troughs with side openings, which distribute the sewage over the bed. The sewage as it goes on the beds shows only a slight milkiness in color, and the solids are not visible to the naked eye, except on closer examination. The ground was graded to proper form, and a grillage of farm tile was laid, leading to a 12-inch main drain pipe, which discharges into the creek. A layer of coarse gravel was placed on the tile drains, and 3 feet of bank gravel, practically unscreened but uniform in quality, completed the filters. A concrete wall surrounds the whole, and gravel walks divide the area into four equal beds.

Bacterial analyses showed over one million bacteria per cubic centimeter in the raw sewage, 40,000 in the effluent from the septic tank, and about 20,000 in the liquid after passing the filter bed. When the thermometer stood at 5 degrees Fahr. below zero, the temperature of the sewage thawed the frozen ground slowly, and at no time was the operation interfered with by freezing, even when the temperature was 20 degrees Fahr. below zero.

XXIII. From a catalog of the Cameron Septic Tank Co., from which the illustration (Fig. 113) of a typical septic tank installation is taken, I quote the following description of the system: —

"The Cameron septic tank system of sewage disposal consists of a tank of suitable dimensions, and so arranged that a mass of putrefactive organisms or anaërobies are developed therein of a character and quantity sufficient to liquefy the solid matter of the flowing sewage. It involves the complete separation of the anaërobic or putrefactive germs from the aërobic or nitrifying organisms, so that the work of both is performed unimpeded by the presence of the other; the septic tank is the workshop of the anaërobies, where ideal conditions are provided for their development and activity, i.e. the absence of air, light, and agitation; while in the contact or filter beds these conditions are completely reversed, and an ideal home for the development and activity of the aërobies is provided. The result accomplished by the Cameron process is the liquefaction and purification of sewage on a practical and efficient scale ' avoiding the formation of sludge.' "

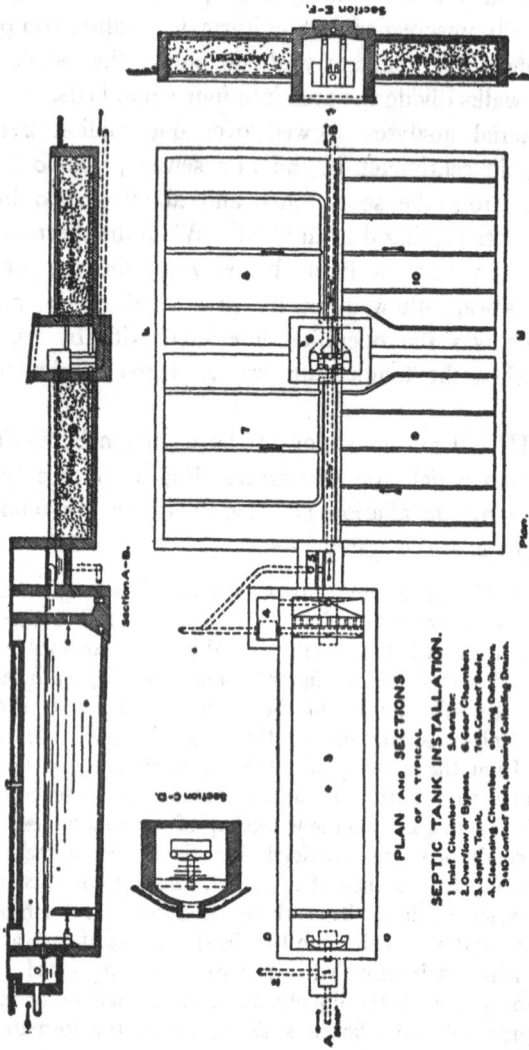

PLAN and SECTIONS
OF A TYPICAL
SEPTIC TANK INSTALLATION.

1. Inlet Chamber   5.Aerator
2.Overflow or Bypass   6.Scar Chamber
3. Septic Tank   7.8.Contact Beds
4.Cleansing Chamber   showing Distribution
9-10 Contact Bods, showing Collecting Drain

FIG. 113.—SEWAGE DISPOSAL BY SEPTIC TANK AND CONTACT BEDS.

" It may be divided into three periods:

1. The septic, liquefying, putrefactive or anaërobic period;
2. Aërating period;
3. Filtering, aërobic, oxidizing or nitrifying period.

"The first of these, or the septic period, involves two stages:

A. The maturing or ripening stage;
B. The liquefying stage.

"The length of time that the maturing or ripening stage will take to develop varies, because it will depend on the character of the sewage to be dealt with and other varying conditions, but with an average sewage under normal conditions, substantial septic action will not be completely established in less than from six to twelve weeks, and during this time there will be a rapid but decreasing accumulation of solids in the tank. When the maturing stage is complete, and septic action established, an equilibrium exists between the incoming solids and the anaërobic bacterial action set up in the tank; this constitutes the liquefying stage, and as a result of this liquefaction practically no more solids accumulate.

"We come now to the second or aërating period. As the liquid effluent leaves the septic tank it is impregnated with gases produced by anaërobic action or putrefaction, and has a slight odor; to release these gases which are inimical to aërobic action, the effluent is exposed to air and light in thin films, and as the gases escape during this exposure or aëration, a corresponding volume of air is absorbed, so that not only are the anaërobies and aërobies entirely separated, but the effluent is put in the best possible condition for the third or final period referred to above.

"The nature of this last operation will depend on the character of the outlet, and degree of purity desired; where the volume of sewage is small compared with the stream or other body of water into which it is discharged, or when a high degree of purification is considered unnecessary, the tank effluent may be discharged without further treatment; when, on the other hand, a higher degree of purification is essential local conditions will determine whether aërobic bacterial contact, sand filtration, or irrigation shall be resorted to, and to what extent.

"Aërobic bacterial contact consists of two or more beds, constructed preferably of concrete, so as to be made watertight, and filled with suitable material, such as coke breeze, cinders, or furnace slag, screened and freed from dust and particles. They are filled alternately, allowed to stand full from one to three hours and then emptied by means of suitable valves; as the sewage leaves the bed the air is drawn down into the interstices of the filtering material, so that it is thoroughly aërated before being again filled. This alternate filling, emptying, and aëration is controlled by an automatic alternating gear, so that the operation is not dependent upon the fidelity and vigilance of an attendant. Sand filtration and irrigation, the other methods of subsequent treatment, are too well known and understood to need explanation.

"It is impossible to lay down exact rules defining the dimensions and proportions of septic tanks, or as to the most desirable method for the subsequent treatment of the tank effluent; these are matters that depend not only on the character of the sewage to be dealt with, but also on the nature of the outlet and local conditions generally For these reasons, each case must be considered independently in order that due benefit may be derived from any natural advantages that may exist locally as affecting the efficiency or economy of an installation."

If I understand the claims of the Cameron system rightly, it is asserted that the septic tank liquefies *all* the solids in the sewage. Experience with such tanks has shown that the claims are by no means fulfilled.

I think it is of importance to point out that Cameron himself considered the septic tank process merely as a preparatory process, or the first stage of purification. The examination of Fig. 113, showing a typical sewage treatment installation according to the Cameron system, makes it evident that he considers the second stage of purification by contact beds, as shown, necessary to obtain good results.

XXIV. The folding *Plate* illustrates part of an elaborate plan prepared by the firm of Waring, Chapman & Farquhar, showing several schemes for the disposal of the sewage for a very large country mansion, and for the stable, conservatory, and other minor buildings attached to the estate. Only the mansion is shown here.

Scheme A proposed a sewage disposal plant by the "Waring" subsurface irrigation system. The sewage tanks are located east of the mansion, at a distance of about 350 feet, and consist of settling tank, flush tank with automatic siphon, and gate chamber to distribute the sewage into either of two fields, each comprising 10 lines of absorption drains, each line being about 150 feet long.

The objection made to this scheme was that the sewage field was located within about 100 feet of the main drive and approach to the mansion, and it was accordingly not carried out.

Scheme B contemplated a similar disposal system, to be located about 550 feet north of the house, but this also was rejected.

Scheme C, which was the one adopted by the owner, illustrates the arrangement of a sewage disposal system by surface irrigation. In this case the tanks are placed 300 feet west of the house, and the disposal is by means of three sections, caoh having ten surface outlets.   The field for sewage irrigation is placed about 400 feet away from and about 100 feet above a brook, and it is expected that the sewage will have entirely soaked into the ground and have been purified before any part of it could reach the brook.

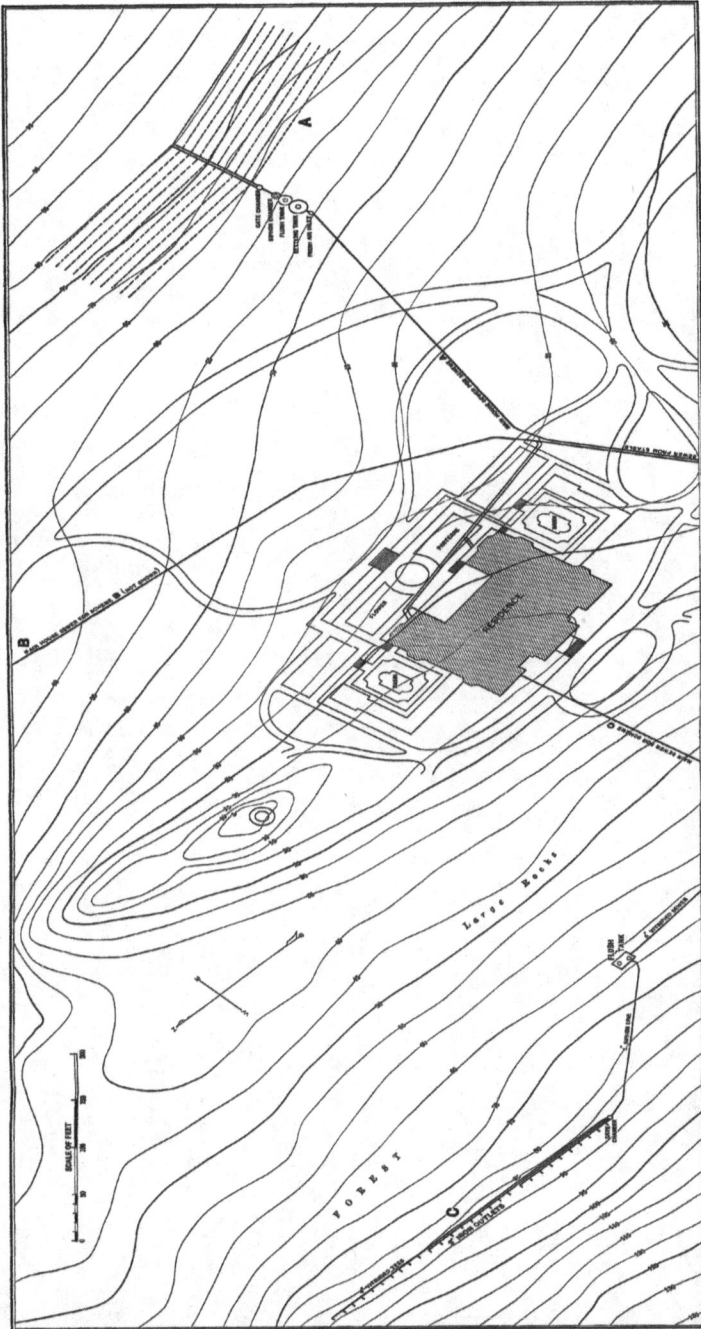

TOPOGRAPHICAL MAP OF PART OF COUNTRY ESTATE, SHOWING PROPOSED SCHEMES FOR SEWAGE DISPOSAL.